W9-CEI-095
Doodle
yourself
smart...
Physics
book

F
b
G
C
a
°S
B
A

Doodle yourself smart...

Physics

book

over 100 doodles and problems to solve!

1	15	14	4
12	6	7	9
8	10	11	5
13	3	2	16

$\log_3(x+2) + \log_3(x-4) = 3$

$\log_3 (x+2)(x-4) = 3$

$x = 7$

$x = -5$ NO

$m_1 \ddot{u}_n = \alpha (v_n + v_{n-1} - 2u_n)$

$m_2 \ddot{v}_n = \alpha (u_n + u_{n+1} - 2v_n)$

$-m_1\omega^2 u = \alpha v(1 + e^{-ika}) - 2\alpha u$

$-m_2\omega^2 v = \alpha u(1 + e^{ika}) - 2\alpha v$

Doodle Yourself Smart...

THUNDER BAY
P·R·E·S·S
San Diego, California

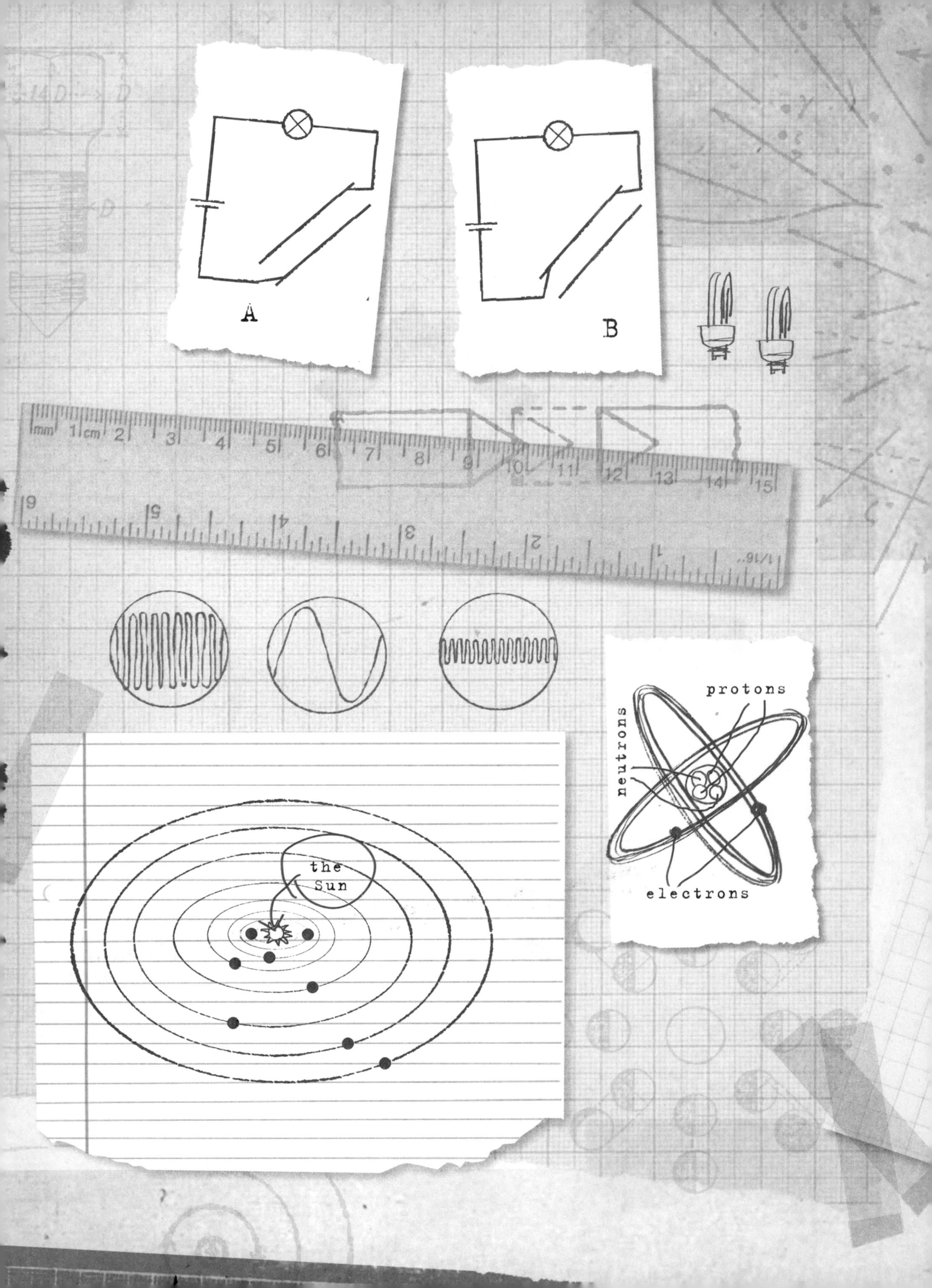

A
B
protons
neutrons
electrons
the
Sun

This book belongs to...
distance in ft
time in seconds
Doodle Yourself Smart...
A
B
C

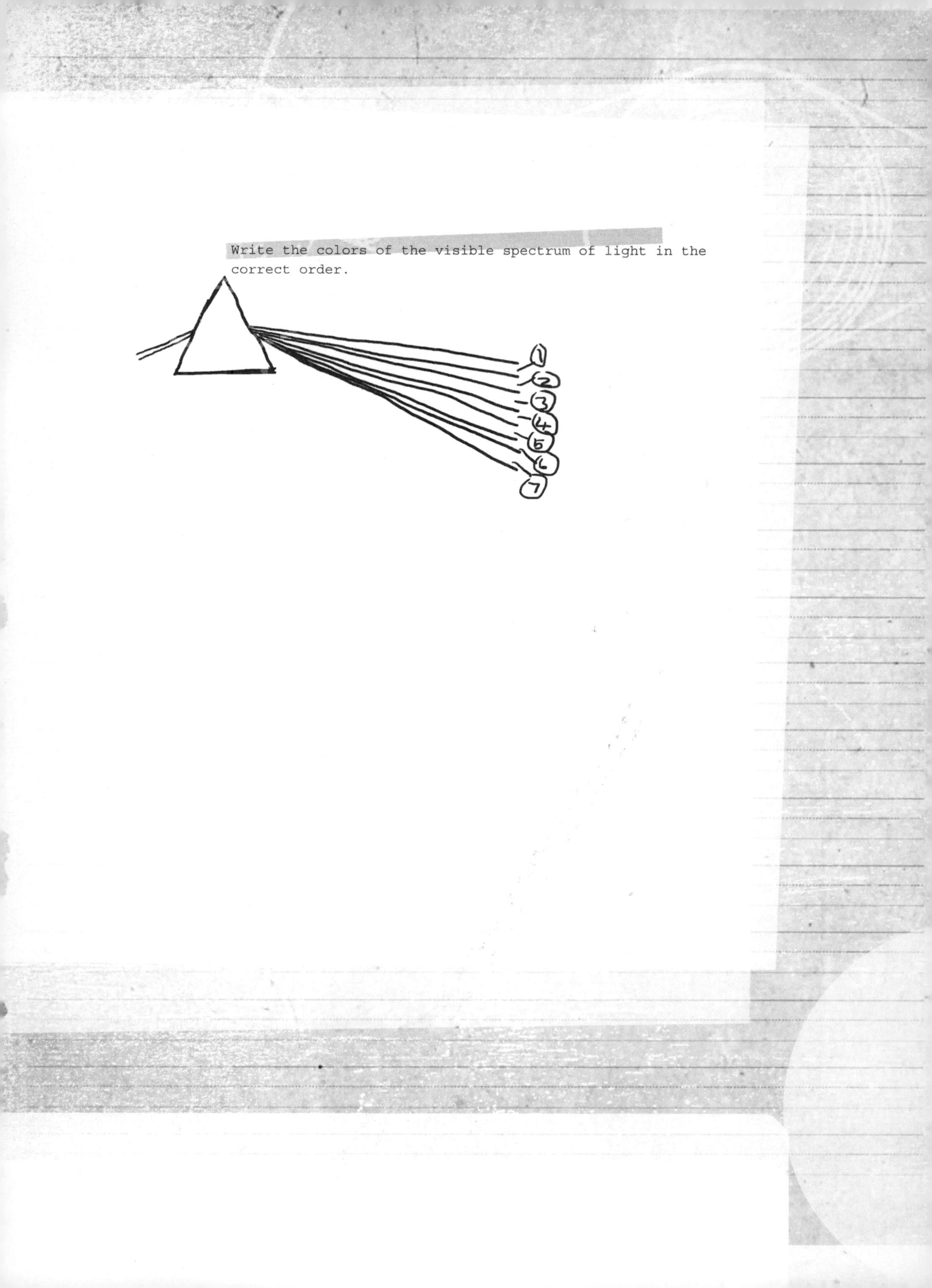
Write the colors of the visible spectrum of light in the correct order.
1
2
3
4
5
6
7

The diagram shows four people standing at different positions on Earth.

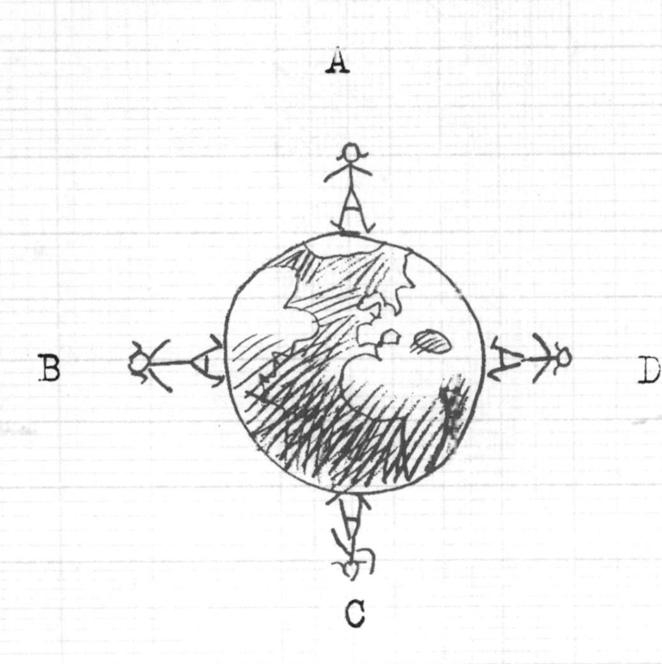

Draw arrows to show the direction of the force of gravity for each person.

answers p120/1

"It is not once nor twice, but times without number that the same ideas make their appearance in the world."

[Aristotle]

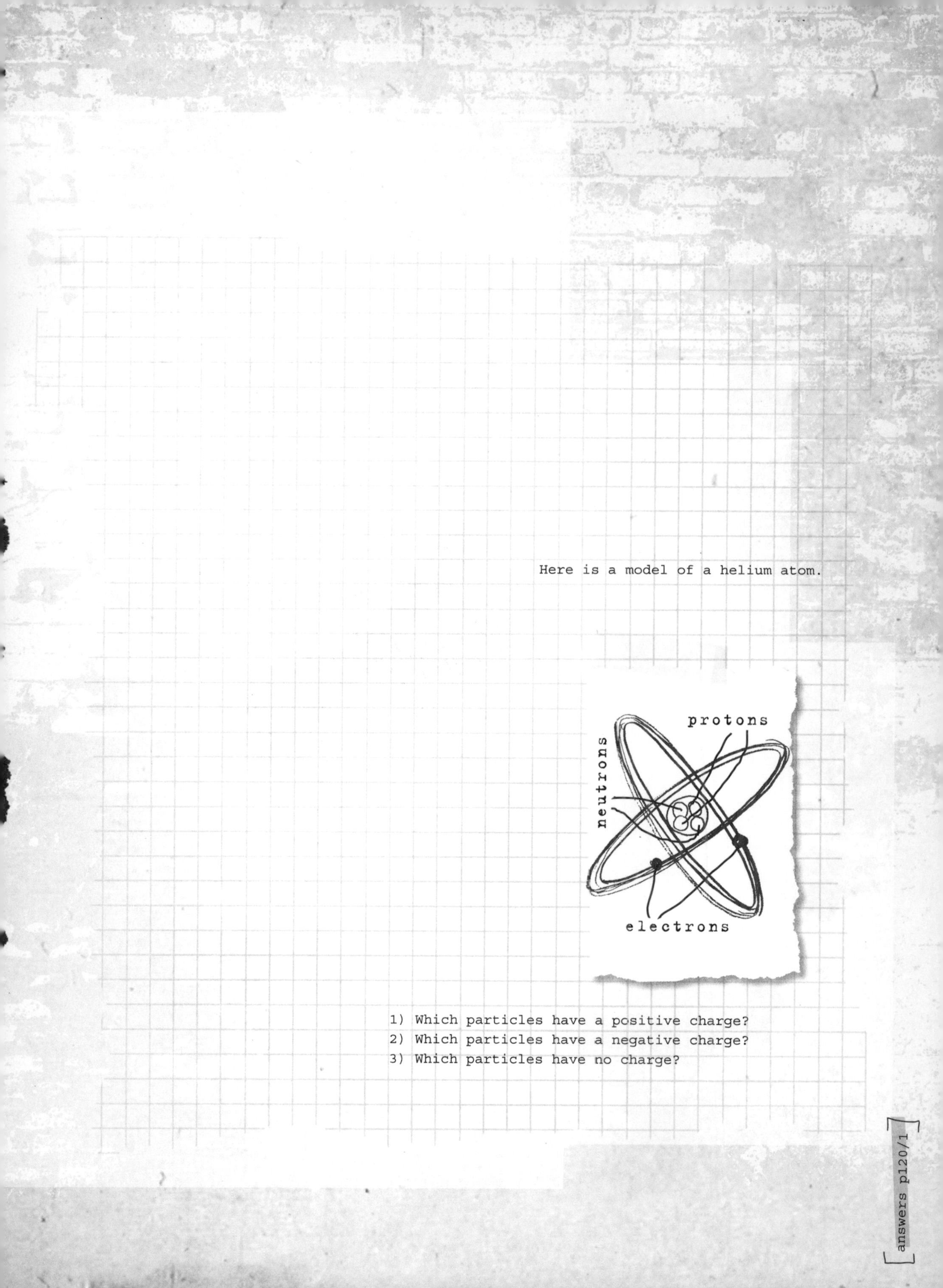

Here is a model of a helium atom.

1) Which particles have a positive charge?
2) Which particles have a negative charge?
3) Which particles have no charge?

answers p120/1

The diagram shows three bar magnets on trolleys.
Magnet B attracts magnet A and repels magnet C.

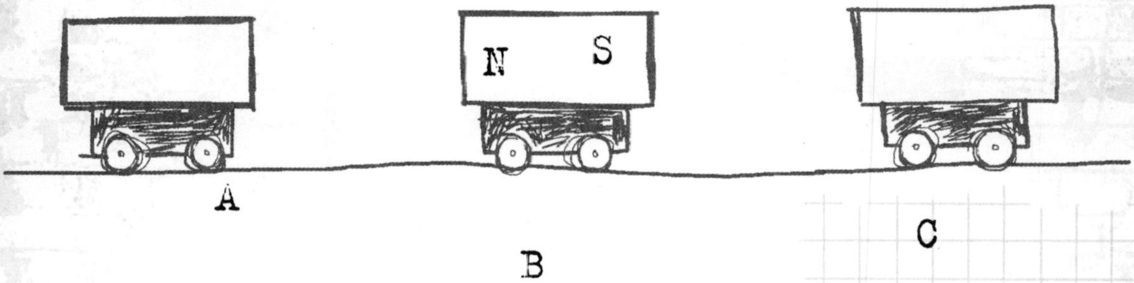

N S

A

B

C

Label the poles on magnets A and C.

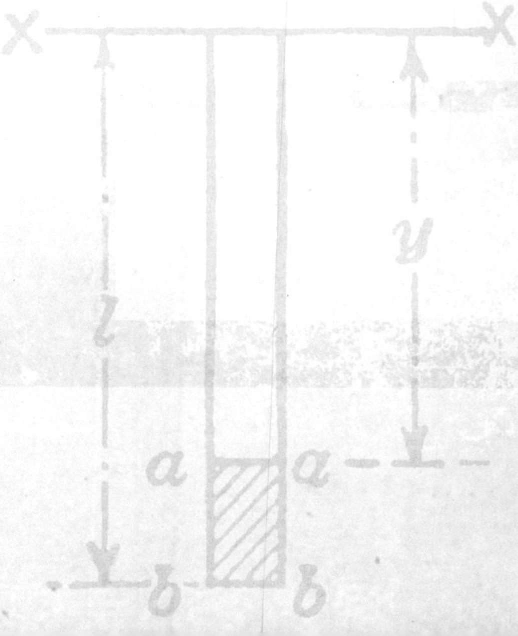

X X

l y

a a

b b

An experiment is set up as shown and then left for half an hour.

What happens to the wooden block?

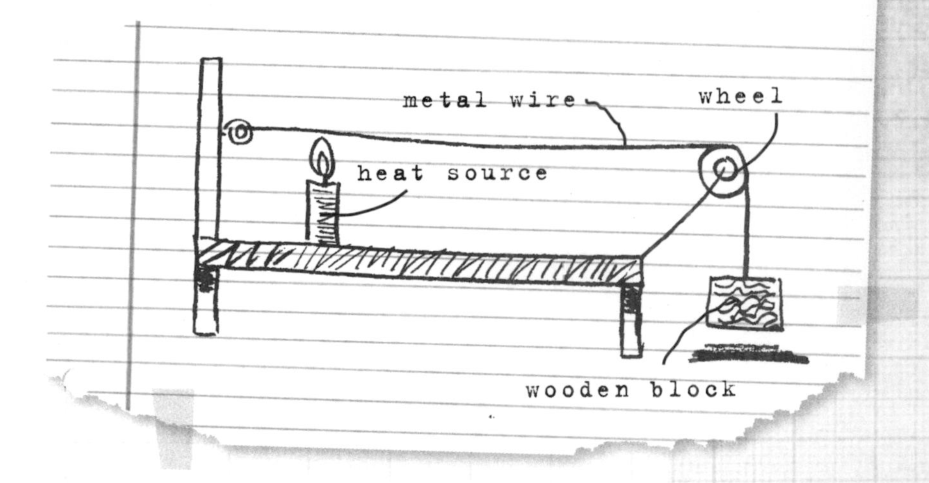

answers p120/1

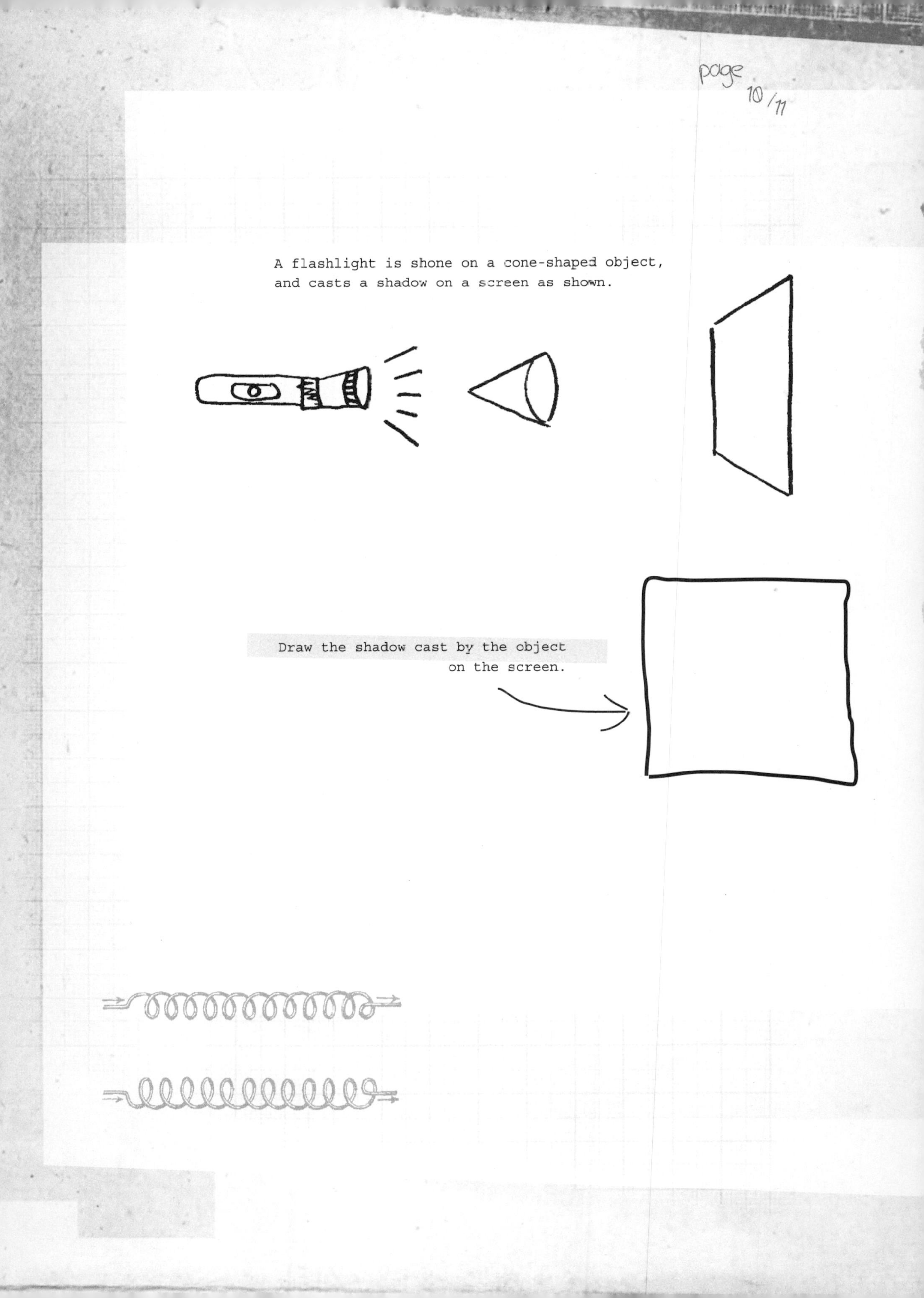
A flashlight is shone on a cone-shaped object,
and casts a shadow on a screen as shown.
Draw the shadow cast by the object
on the screen.

"Measure what can be measured, and make measurable what cannot be measured."

[Galileo Galilei]

Label the planets in their orbits around the Sun.

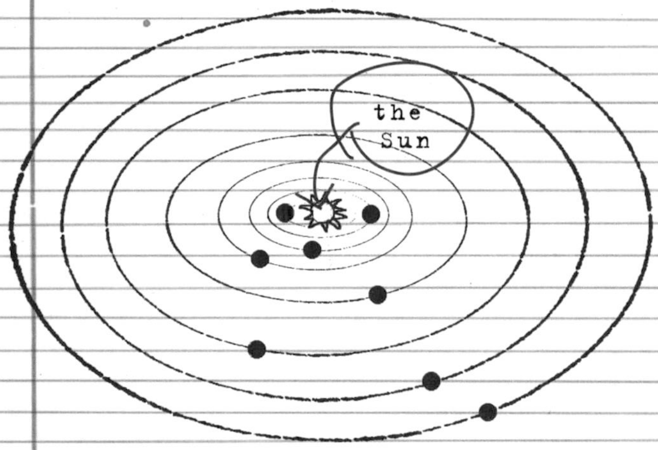

the Sun

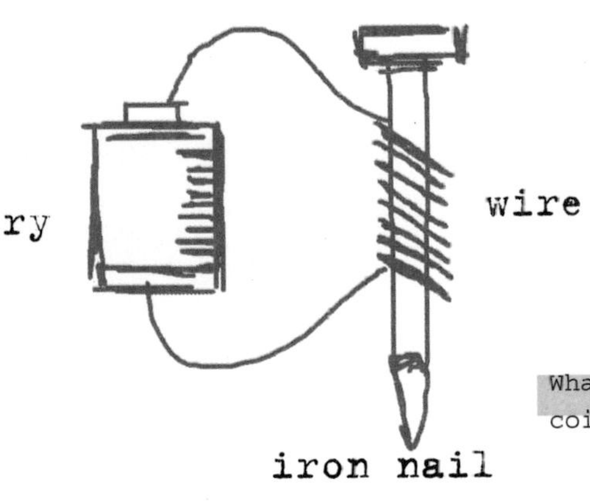

An electromagnet is made by placing a coil of wire around an iron nail and connecting it to a battery. Set up as shown, the electromagnet attracted five steel paperclips.

What would happen if the number of coils in the wire was increased?

answers p120/1

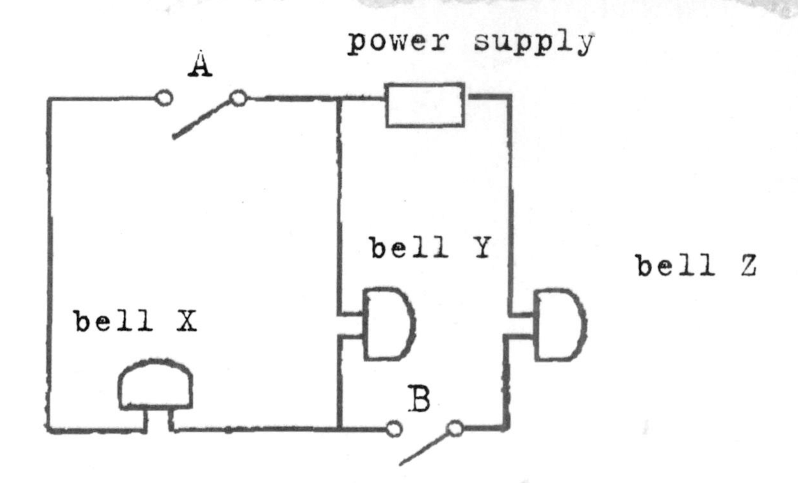

Here is a circuit with two switches and three bells.

1) Which bells will ring if switches A and B are both closed?
2) Which bells will ring if only switch B is closed?

"Every body continues in its state of rest or uniform motion in a straight line, except insofar as it doesn't."

[Sir Arthur Eddington]

The graph shows the speeds and distance from the Sun of the eight planets in the Solar System.

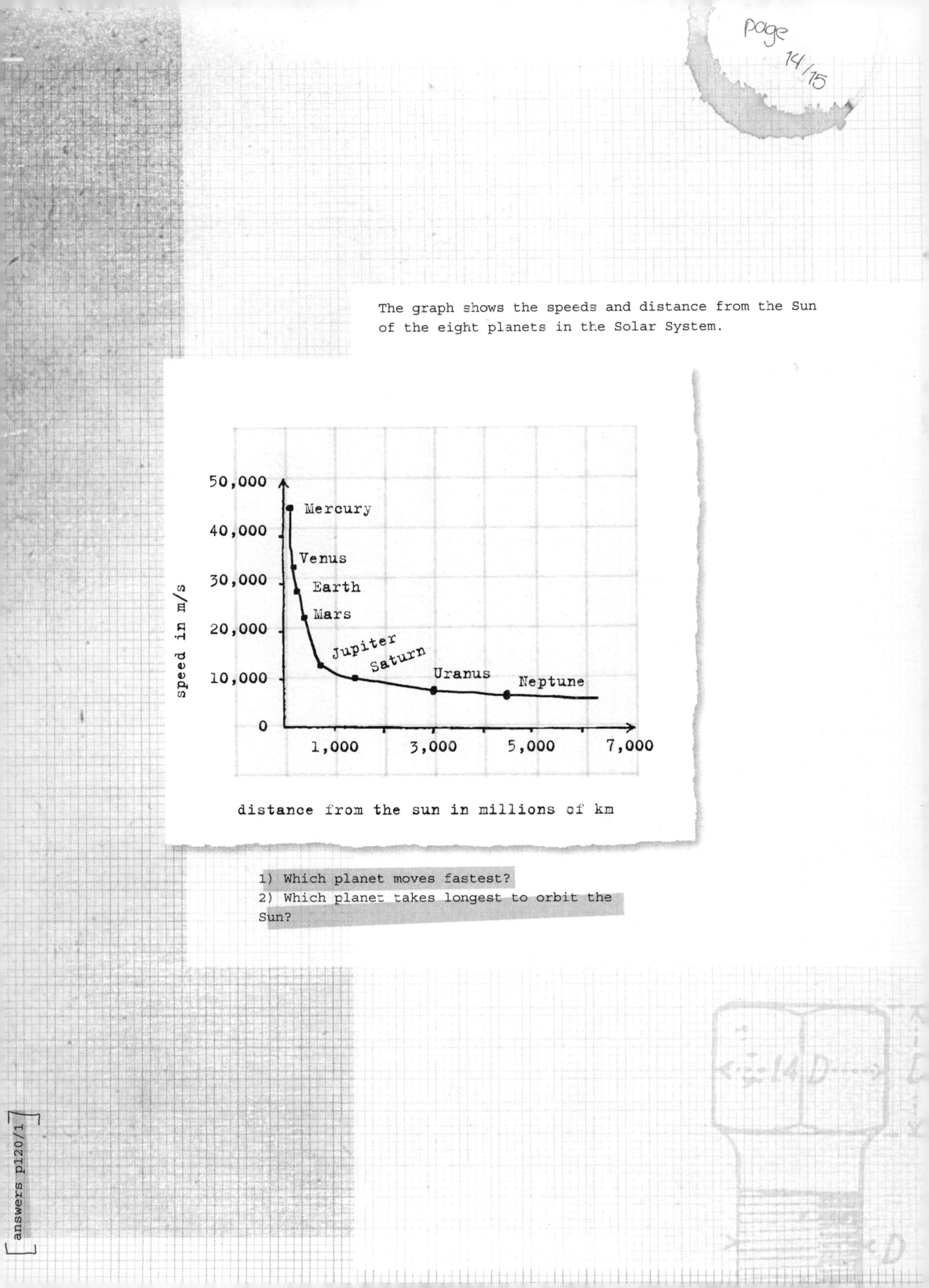

1) Which planet moves fastest?

2) Which planet takes longest to orbit the Sun?

answers p120/1

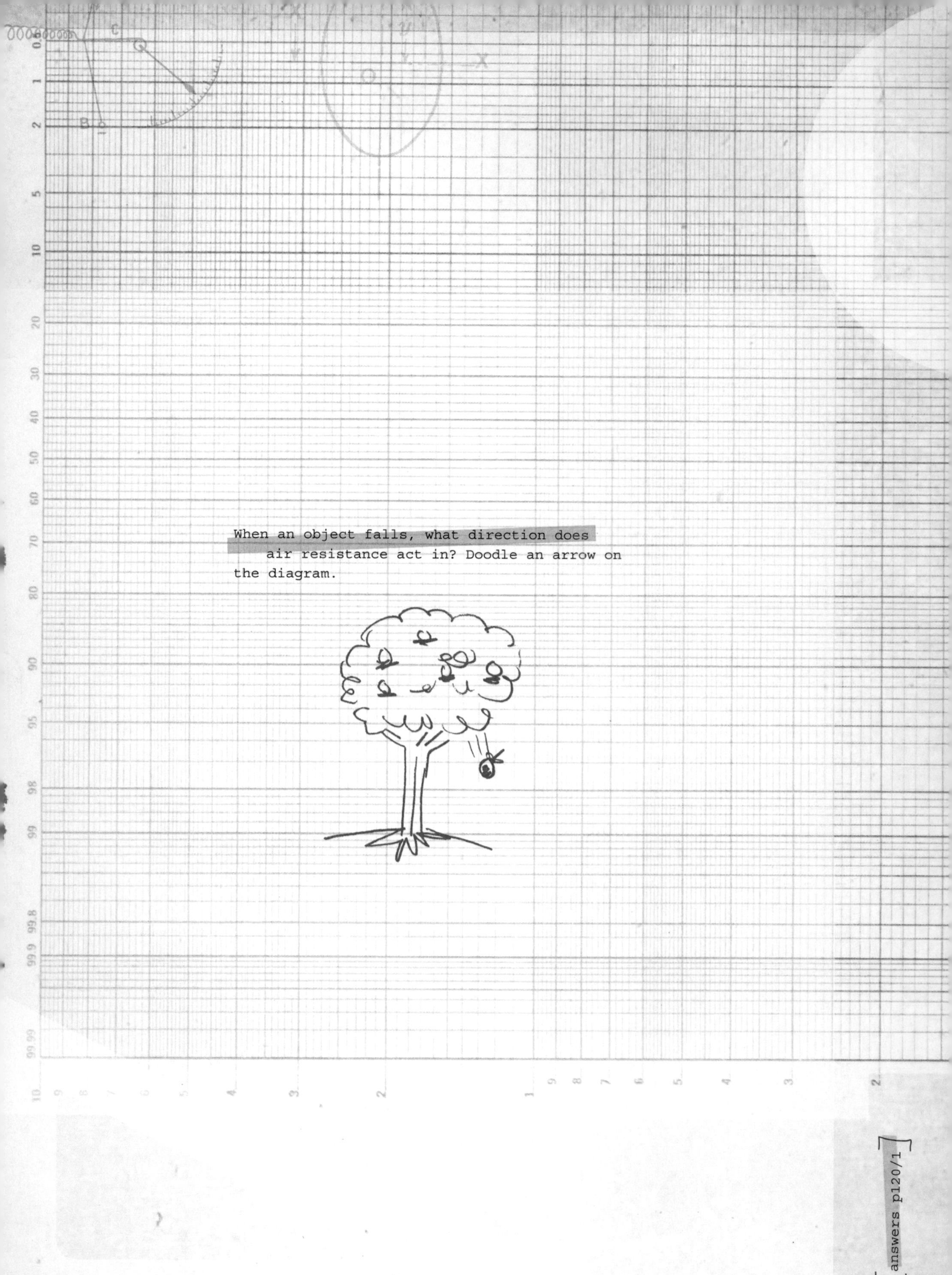

When an object falls, what direction does air resistance act in? Doodle an arrow on the diagram.

[answers p120/1]

page 16/17

An experiment is set up to measure the amount of light that passes through four different liquids: water, cooking oil, blackcurrant squash, and thick black coffee.

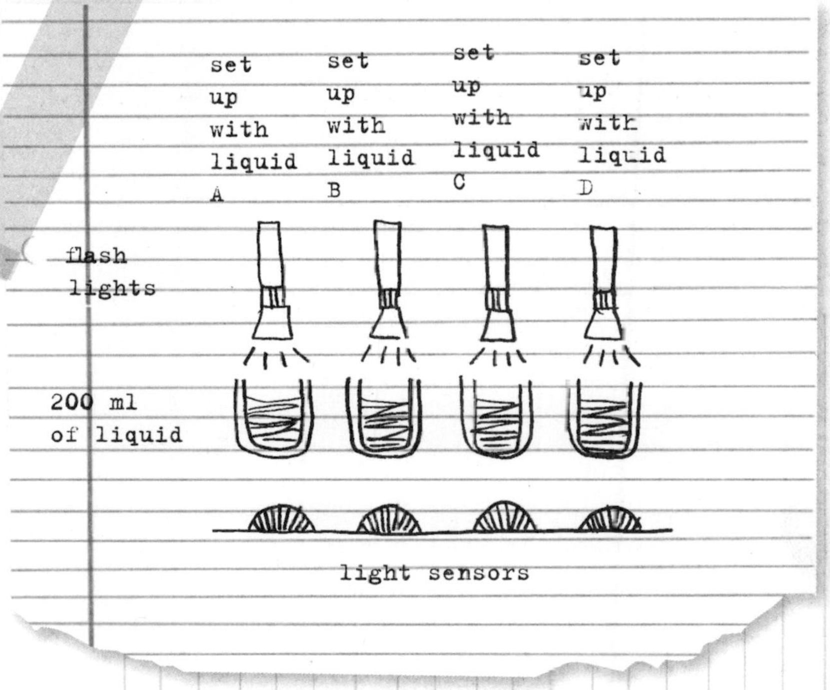

set up with liquid A

set up with liquid B

set up with liquid C

set up with liquid D

flash lights

200 ml of liquid

light sensors

The results are shown in the table.

Liquid	Units of light recorded
A	40
B	100
C	250
D	0

Which liquid is:

Water

Cooking oil

Blackcurrant squash

Thick black coffee

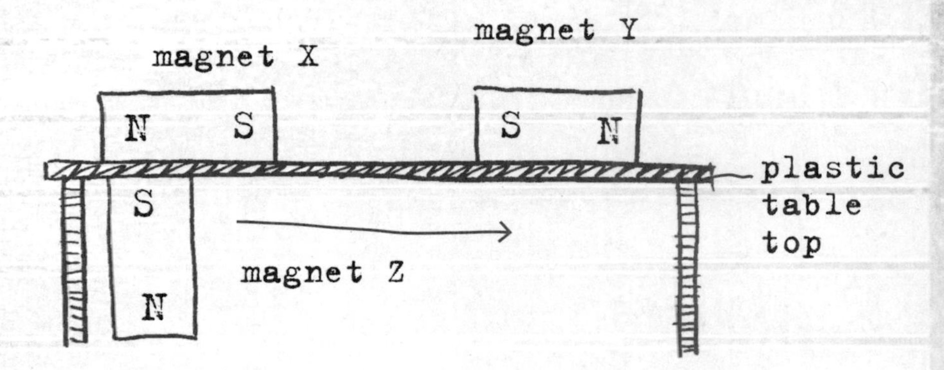

Three magnets are set up as shown. What happens to magnet Y if magnet Z is moved in the direction of the arrow?

"We have no right to assume that any physical laws exist, or if they have existed up until now, that they will continue to exist in a similar manner in the future."

[Max Planck]

On the diagram, draw a ray of light to show how the cyclist can see the motorcyclist.

answers p120/1

An object was placed on a table and lit from different angles.

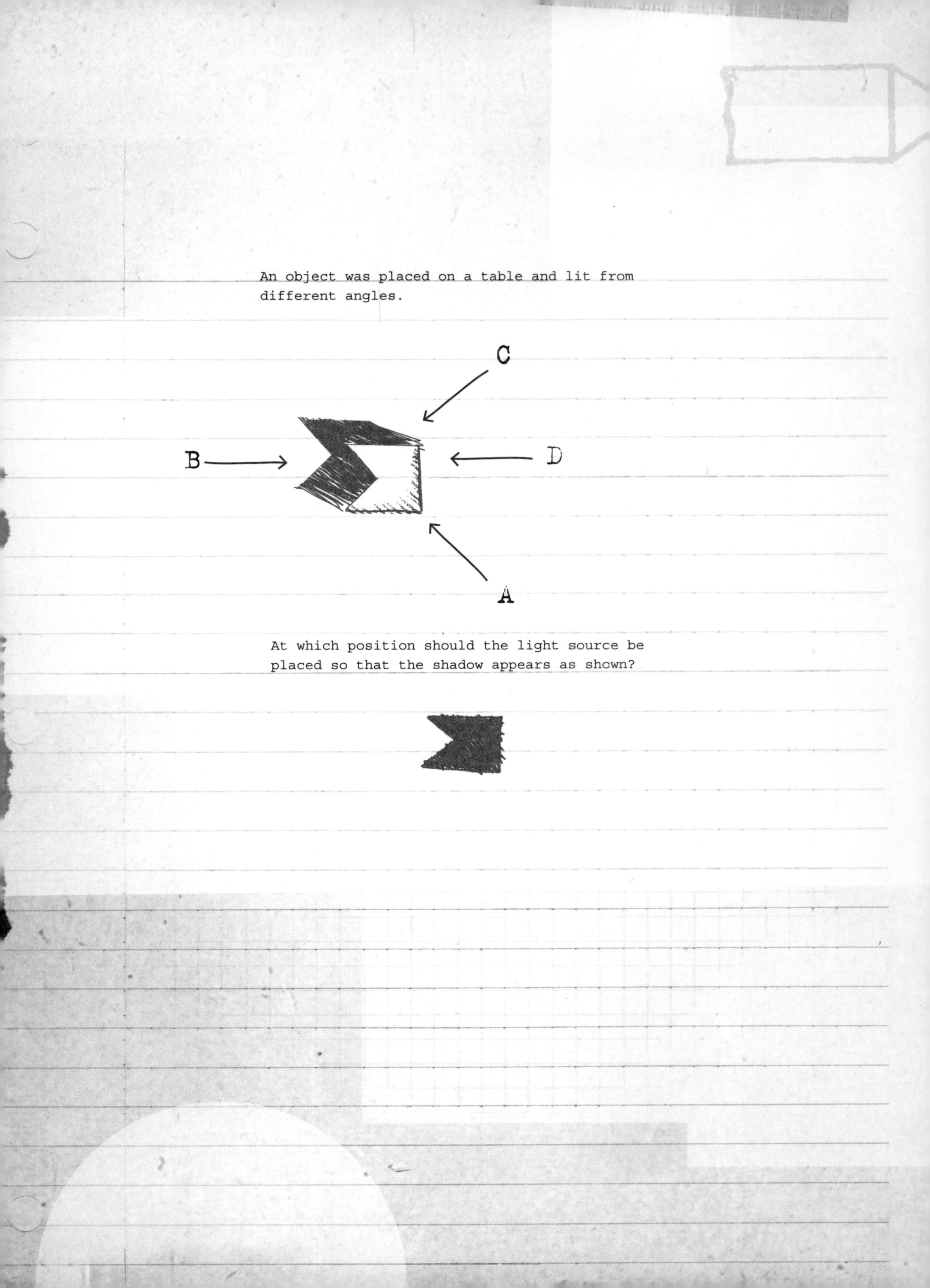

At which position should the light source be placed so that the shadow appears as shown?

An asteroid orbits the sun in the elliptical path shown. At which point, A, B, C, or D, is the effect of the Sun's gravity on the asteroid greatest?

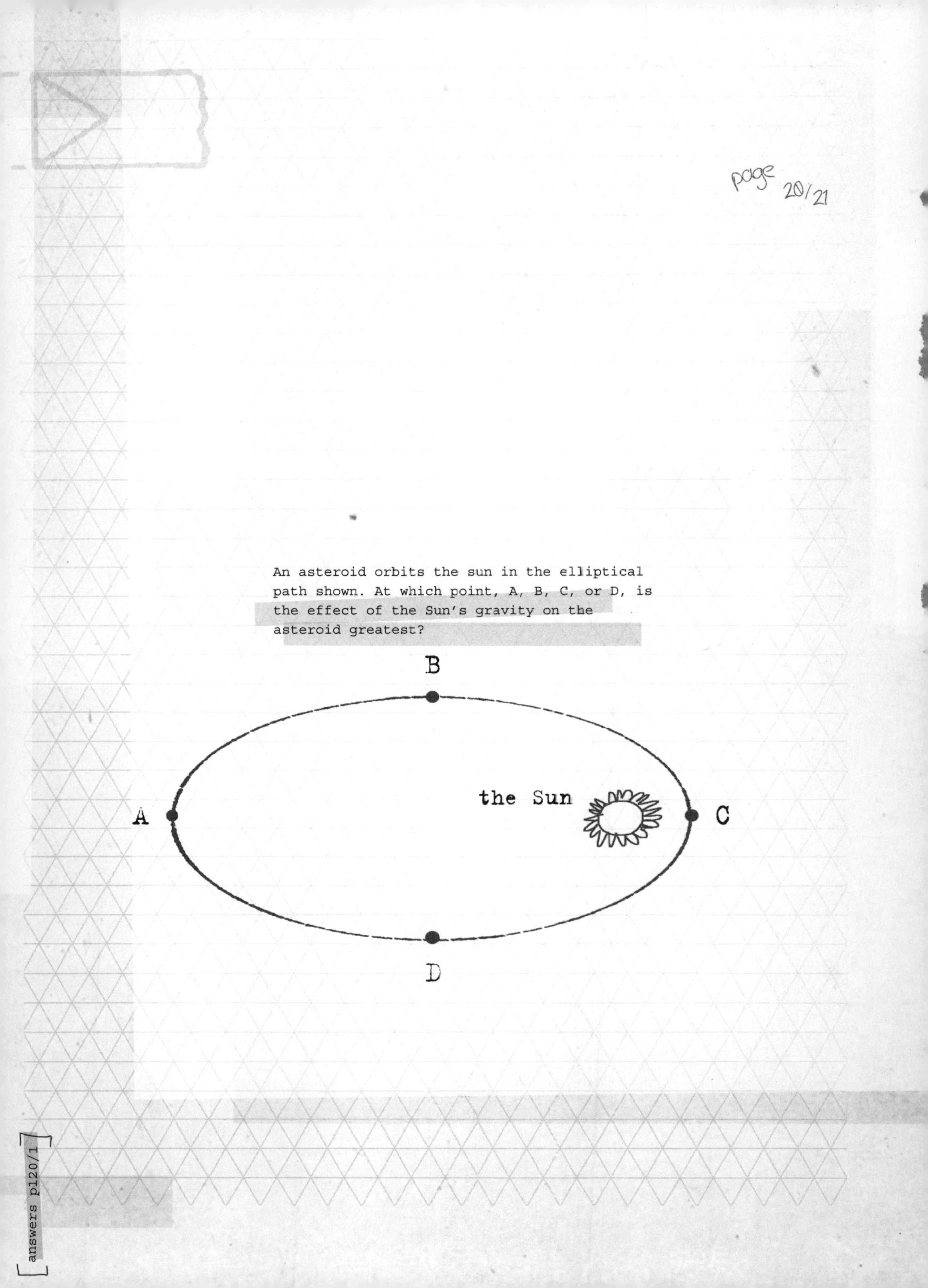

answers p120/1

The strength of four magnets was compared by bringing each of them close to a pile of paperclips and recording how many paperclips each magnet attracted. The results are shown in the table.

Magnet	Distance between magnet and paperclips (inches)	No. of paperclips attracted
A	6	10
B	4	11
C	3	13
D	4	12

Which of the following statements is true?

1) Magnet B is stronger than magnet A.
2) Magnet D is stronger than magnet B.
3) Magnet C is the strongest.

"Life is like riding a bicycle; in order to keep your balance, you must keep moving."

[Albert Einstein]

answers p120/1

An astronaut on the Moon drops a wrench and a feather at the same time from the same height. Which hits the lunar surface first?

Three shapes are cut out of the same-size squares of three different materials: clear plastic, cardboard, and glass.

hole
clear plastic
hole
cardboard
glass
hole

P Q R

The three squares are placed together in a straight line and a flashlight is shone on them.

P Q R

Draw the shape of the shadow on the screen.

The diagrams show two ways of connecting the circuit of a heating element.

Name the two types of circuit.

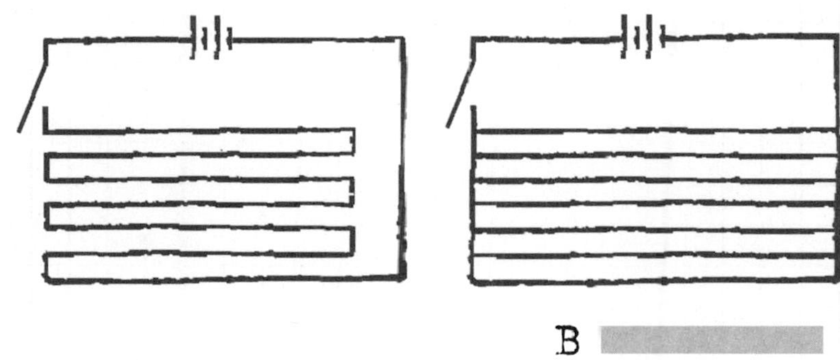

B

A

answers p120/1

Five magnets are stacked as shown.
If A is a north pole, what is J?

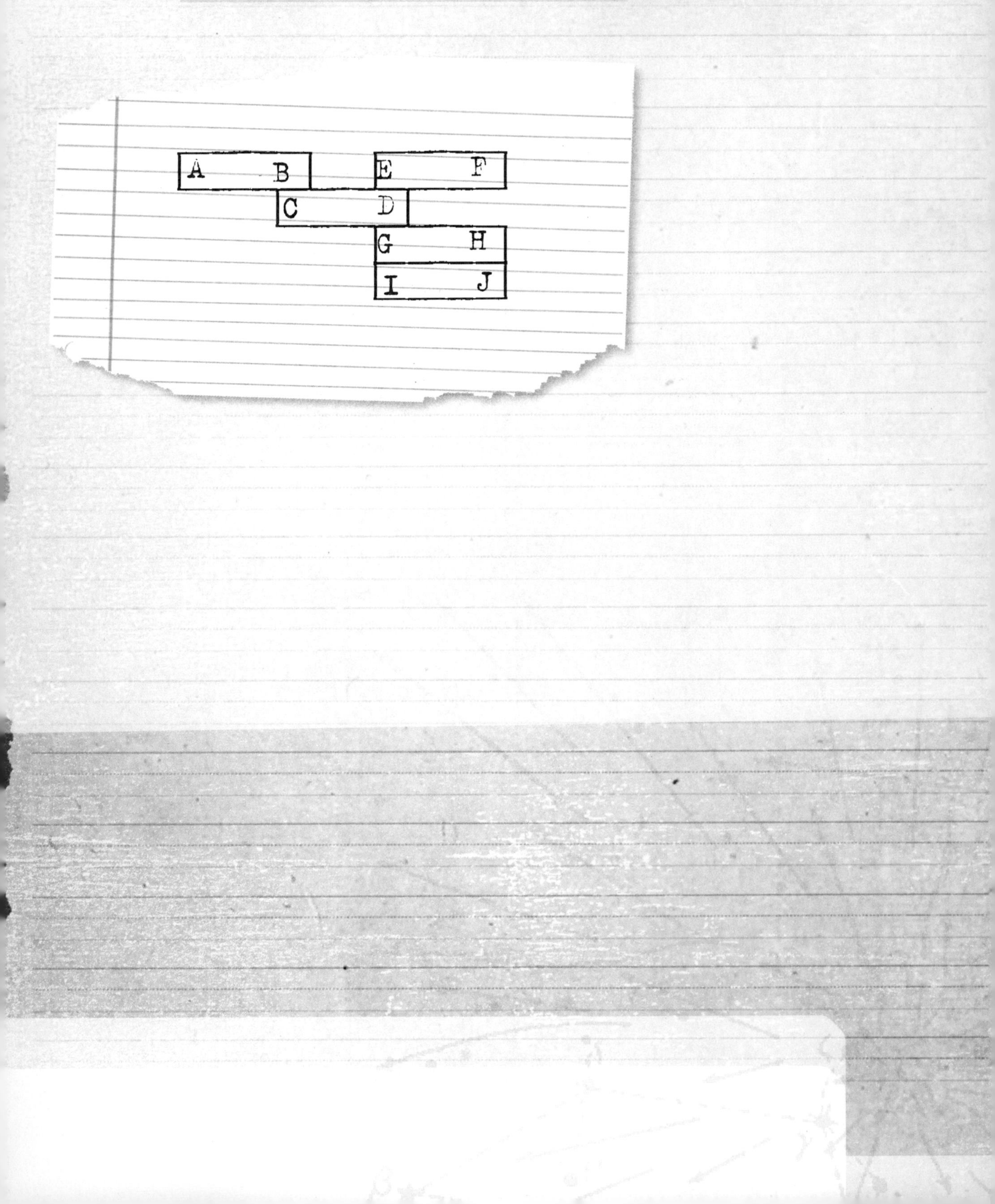

page 26/27

You are standing still on a tennis court, holding your tennis racket. A tennis ball is coming toward you from your opponent.

Which two forces are balanced in this scenario?

A) Air resistance on the ball and gravity
B) The force applied to the ball by your opponent and air resistance
C) Your weight and the reaction force from the ground

answers p120/1

"I can calculate the motion of heavenly bodies, but not the madness of people."

[Sir Isaac Newton]

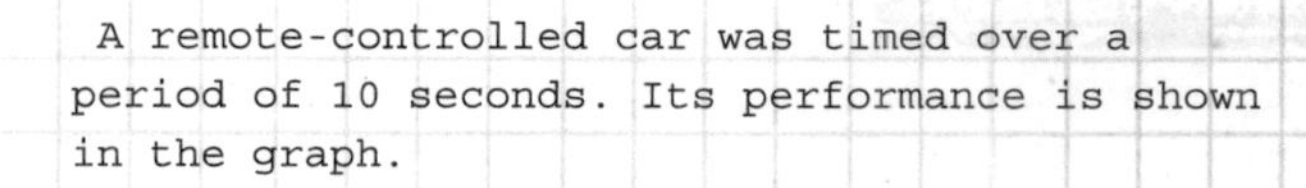

A remote-controlled car was timed over a period of 10 seconds. Its performance is shown in the graph.

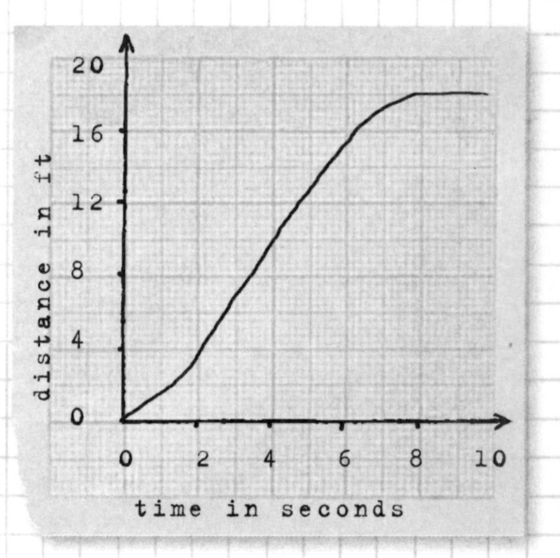

What was the average speed of the car between 0 and 10 seconds?

page 28/29

Link the objects to the type of energy they contain.

Roller coaster at the top of the ride	**Chemical energy**
Vibrating guitar string	**Thermal energy**
Cup of coffee	**Kinetic energy**
Battery	**Gravitational potential energy**
Parachutist in freefall	**Sound energy**

answers p120/1

Which of the three spring balances will show
the highest reading?

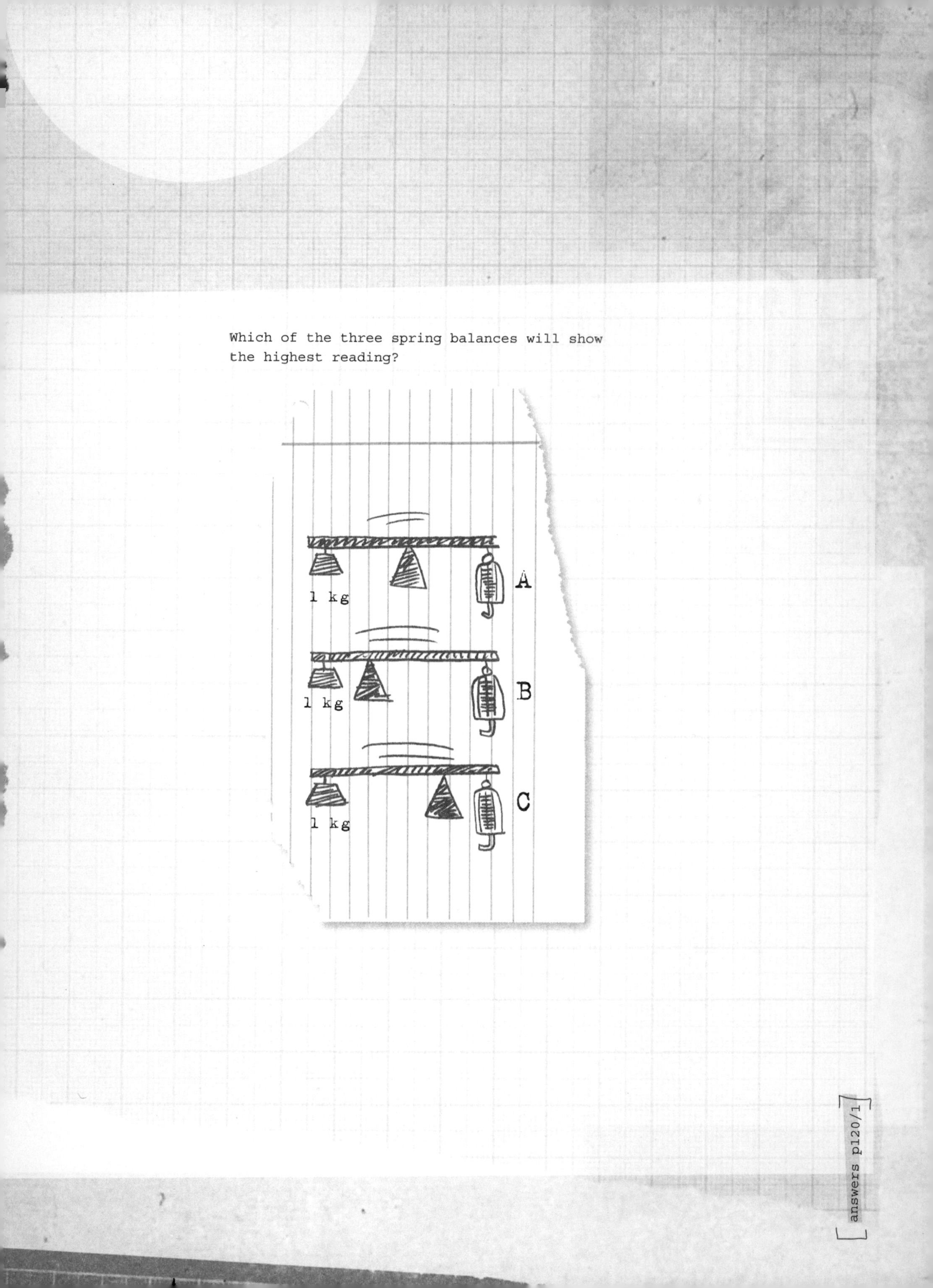

answers p120/1

page 30/31

For a light over a staircase, there is a two-way switch at the bottom and another at the top. In each of these diagrams, is the light on or off?

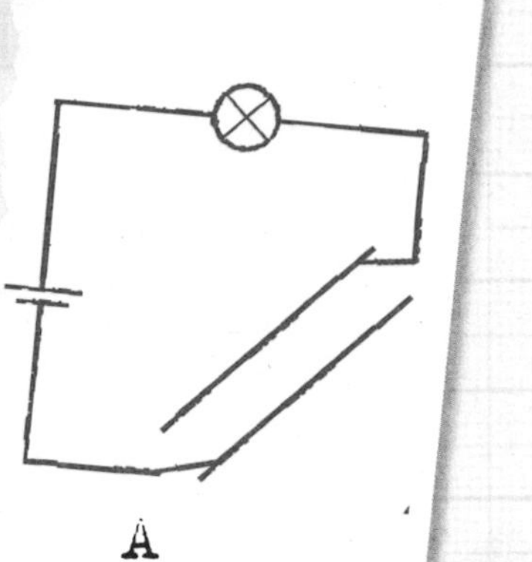

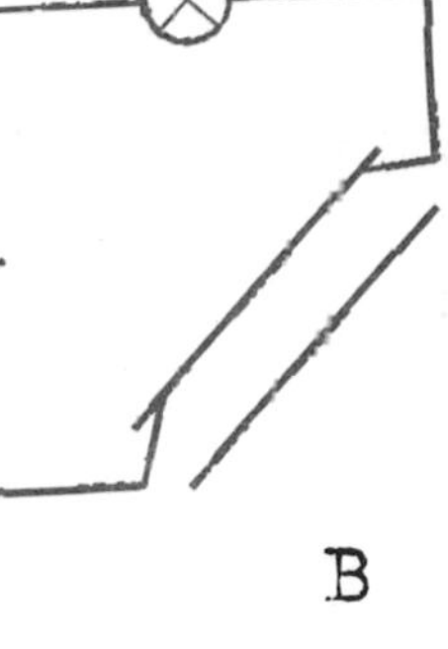

"All science is either physics or stamp collecting."

[Ernest Rutherford]

In a wind turbine, what kind of energy is transferred from the wind to the blades?

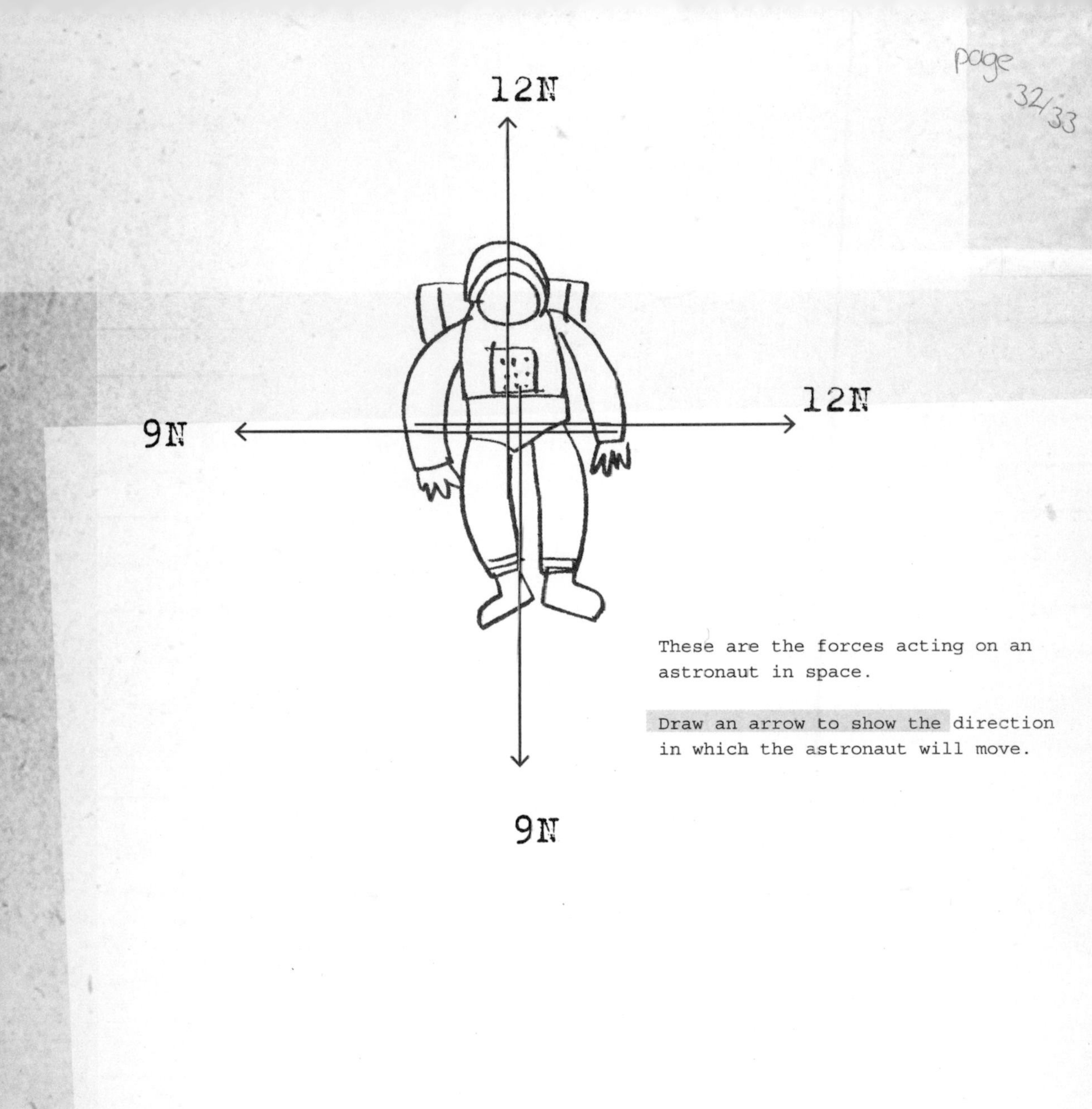

These are the forces acting on an astronaut in space.

Draw an arrow to show the direction in which the astronaut will move.

answers p120/1

"I am undecided whether or not the Milky Way is but one of countless others all of which form an entire system. Perhaps the light from these infinitely distant galaxies is so faint that we cannot see them."

[Johann Lambert]

The diagram shows a circuit for a front lamp and a rear lamp on a bicycle. Both lamps are connected to the same battery.

On the diagram, mark an X to show the position of a switch to turn on only the front lamp. Mark a Y to show the position of a switch to turn both lamps on and off at the same time.

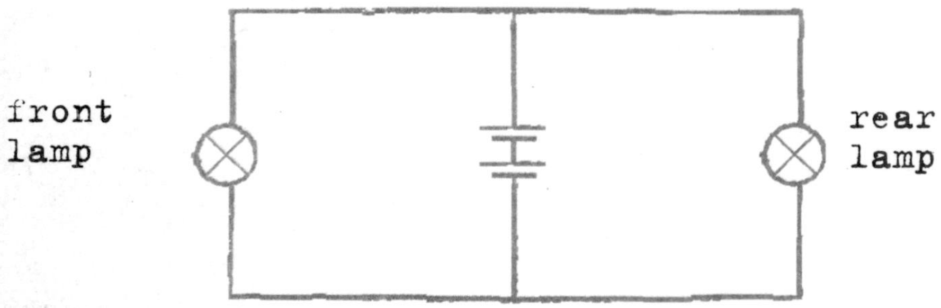

front lamp

rear lamp

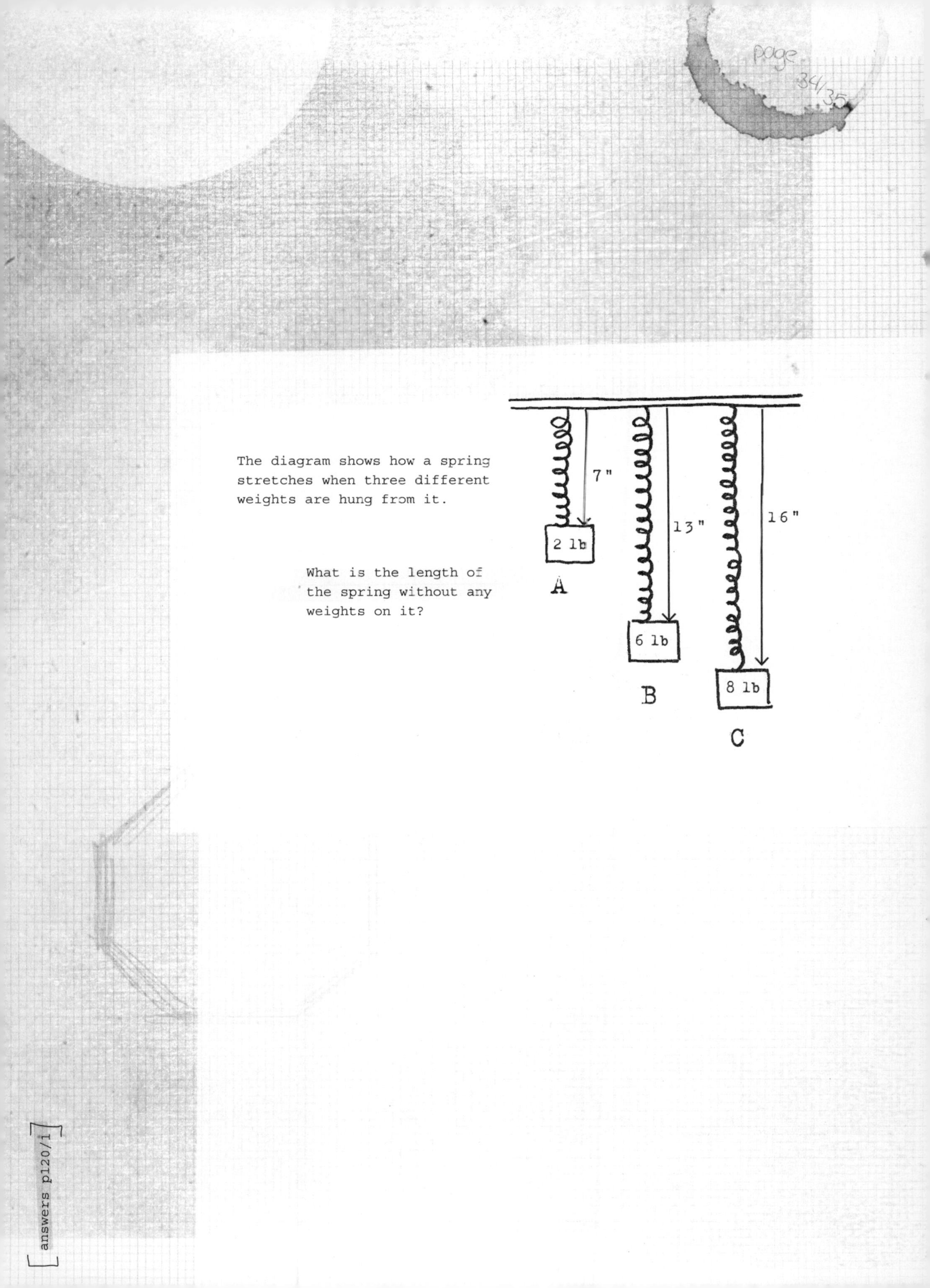
page 34/35
The diagram shows how a spring stretches when three different weights are hung from it.
What is the length of the spring without any weights on it?
7"
2 lb
A
13"
6 lb
B
16"
8 lb
C
answers p120/1

Two gears, A and B, are connected by a chain.
Gear A has 6 teeth and gear B has 12 teeth.

1) If gear A turns in the direction shown, in which direction will gear B turn?
2) How many turns will gear B make if gear A makes 8 turns?

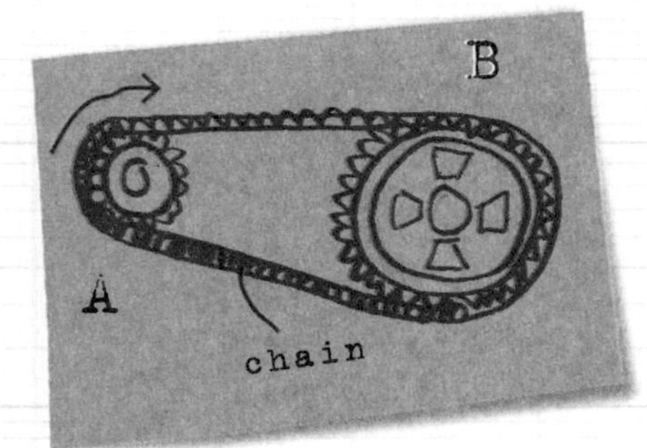

Draw a circuit diagram to represent the electrical circuit below.

answers p120/1

The bar chart shows the force of gravity on the eight planets in the Solar System. On which planet would a spaceship need the most force to take off?

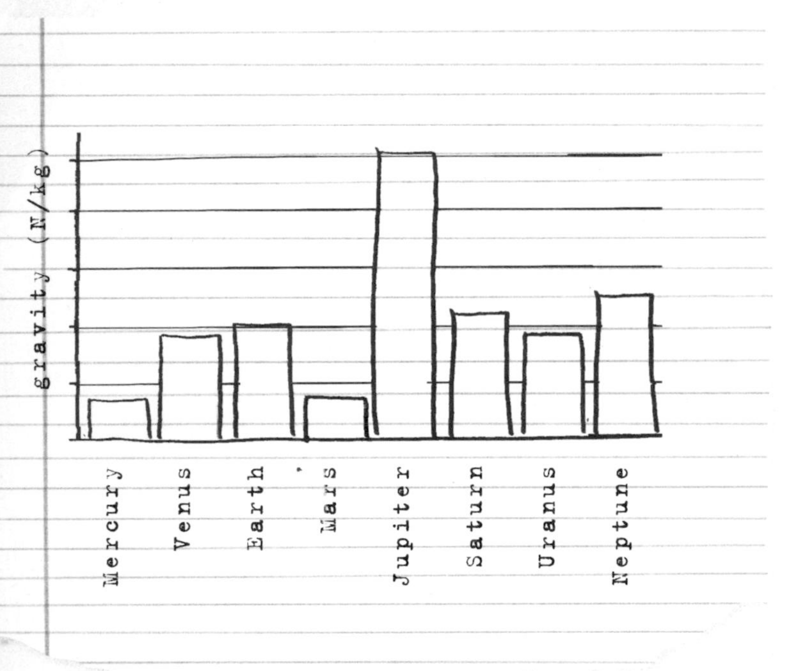

answers p120/1

"Everything existing in the Universe is the fruit of chance and necessity."

[Democritus]

A fencepost in the ground is 1.6 meters tall. A hole is bored in it half way up and a rope attached.

If the rope is pulled with a force of 300 newtons, what is the turning moment about the pivot point P?

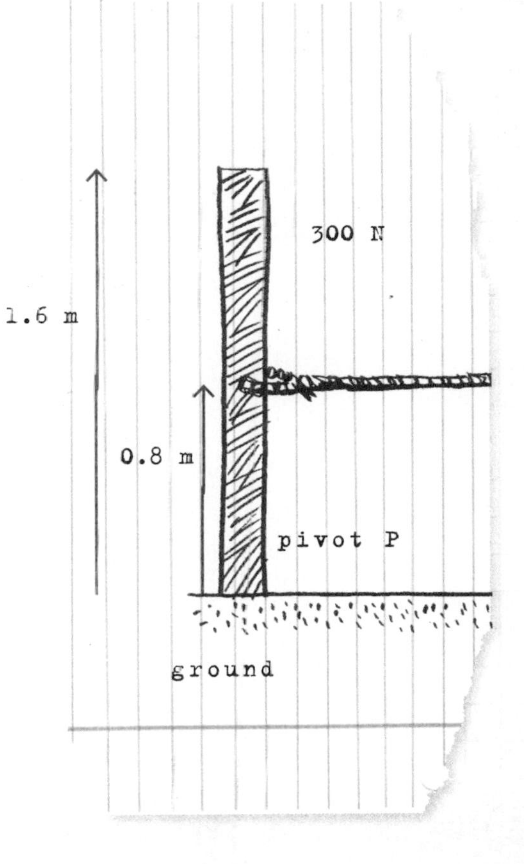

The diagram shows the path of a comet orbiting the Sun.

Mark the position at which the comet is traveling fastest.

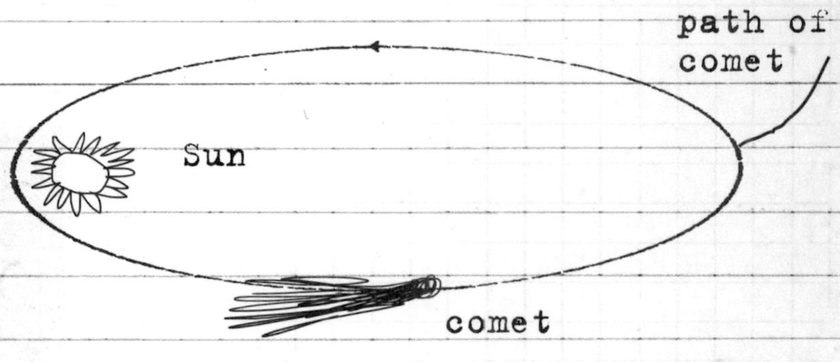

path of comet

Sun

comet

The diagram shows a ray hitting a plane mirror. Draw the reflected ray on the diagram.

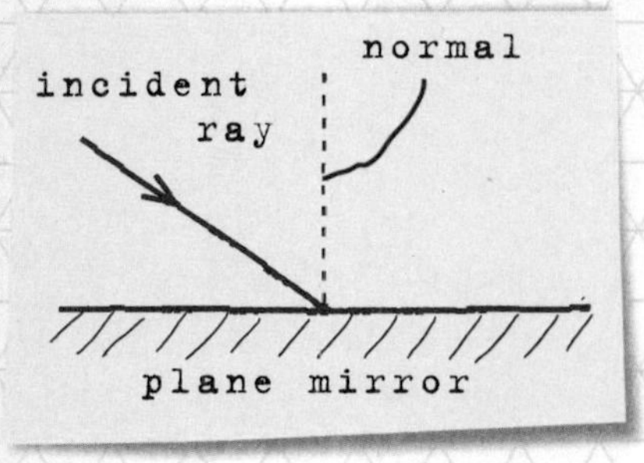

incident ray

normal

plane mirror

answers p122/3

You are given three batteries, two lamps, and some electrical wire.

In what arrangement of the batteries, series or parallel, would the lamps be dimmer? Doodle a diagram of your circuit.

"What we know is not much. What we do not know is immense."

[Pierre Simon Laplace]

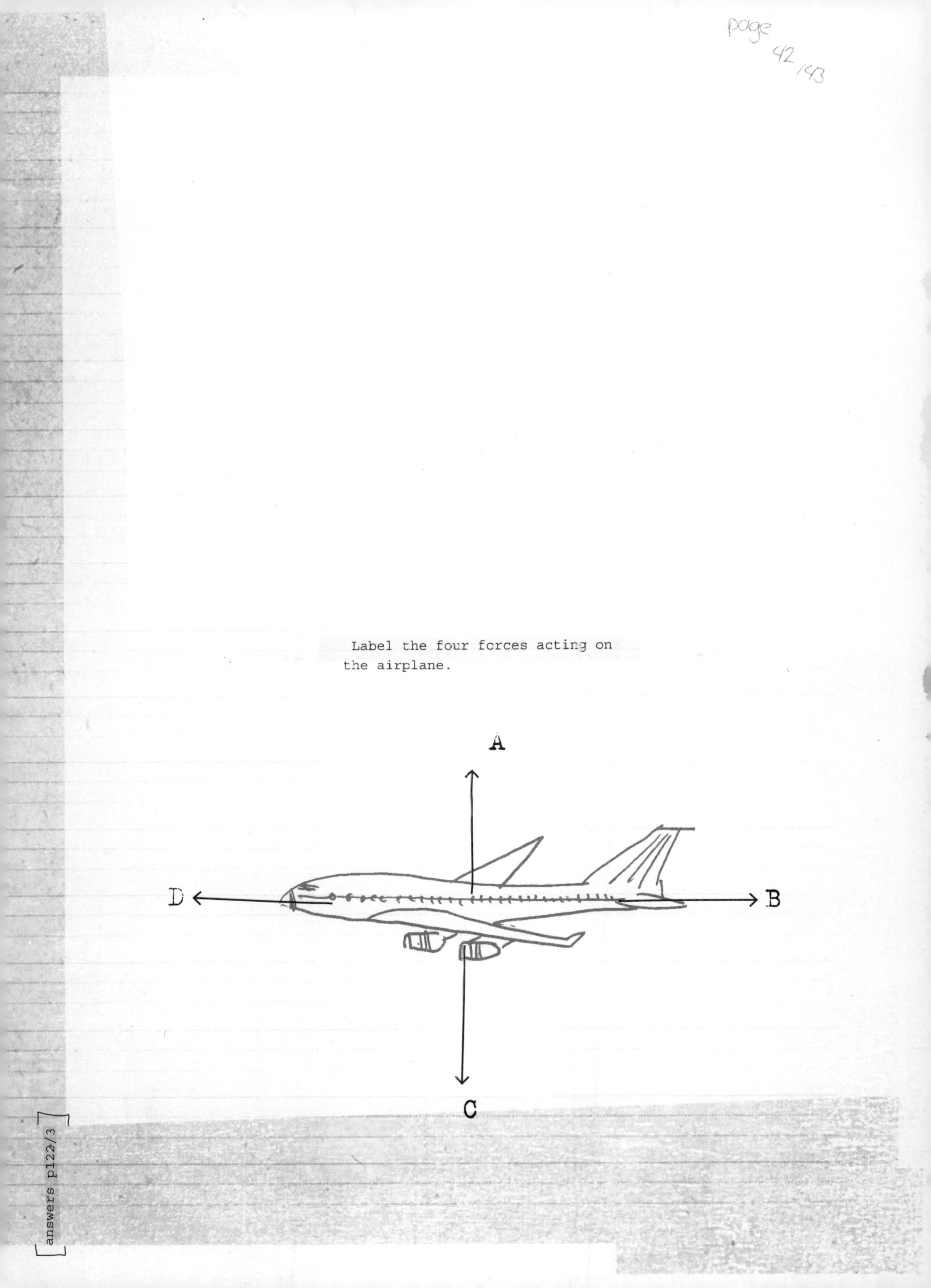
Label the four forces acting on
the airplane.
A
D
B
C
answers p122/3
page 42/143

In which direction does the Earth spin?

A) North to south
B) East to west
C) West to east
D) South to north

answers p122/3

A circuit is set up as shown to test the conductivity of four rods of different materials: W, X, Y, and Z.

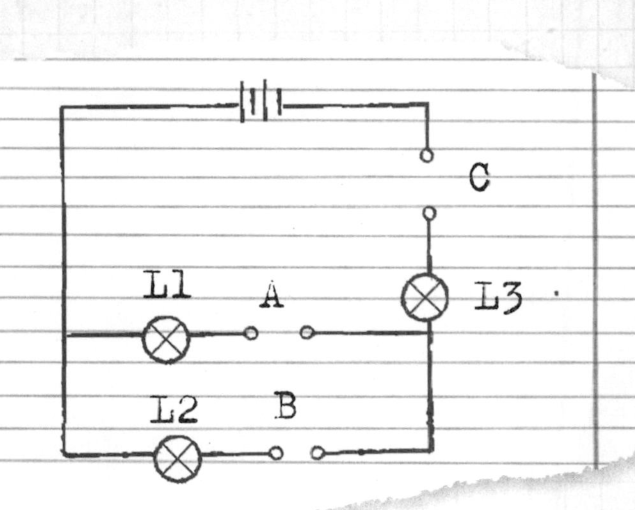

Each of the four rods are placed in the circuit at points A, B, and C. The results are shown in the table.

A	B	C	Lamp 1	Lamp 2	Lamp 3
X	Y	W	Not lit	Not lit	Not lit
X	W	Y	Lit	Not lit	Lit
W	Y	Z	Not lit	Not lit	Not lit
Z	Y	X	Not lit	Lit	Lit

Which rods conduct electricity?

"We can't solve problems by using the same kind of thinking we used when we created them."

[Albert Einstein]

Label the wave diagram using the words below.

wavelength crest amplitude trough displacement

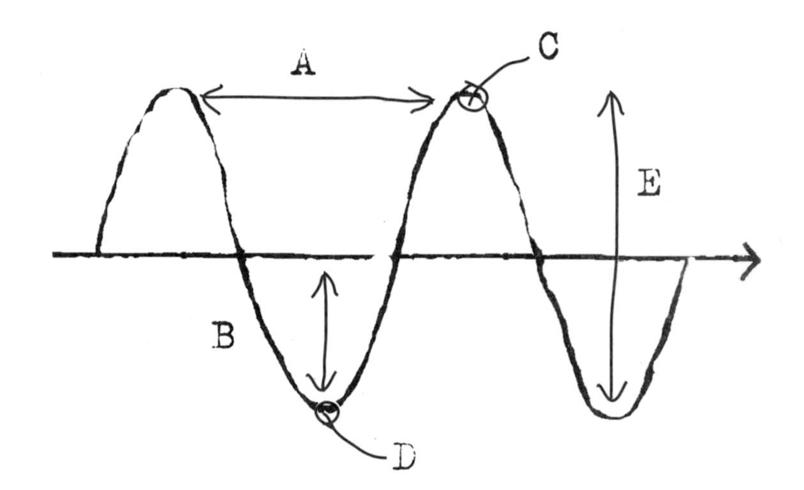

The diagram shows the transfer of energy for a hydroelectric scheme.

Complete the types of energy in the transfers.

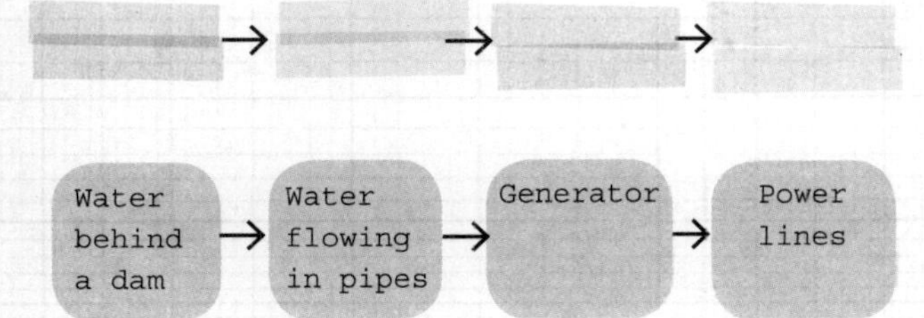

answers p122/3

The graph shows the upward force and the weight of a space shuttle. What is the resultant force on the shuttle at 20 seconds?

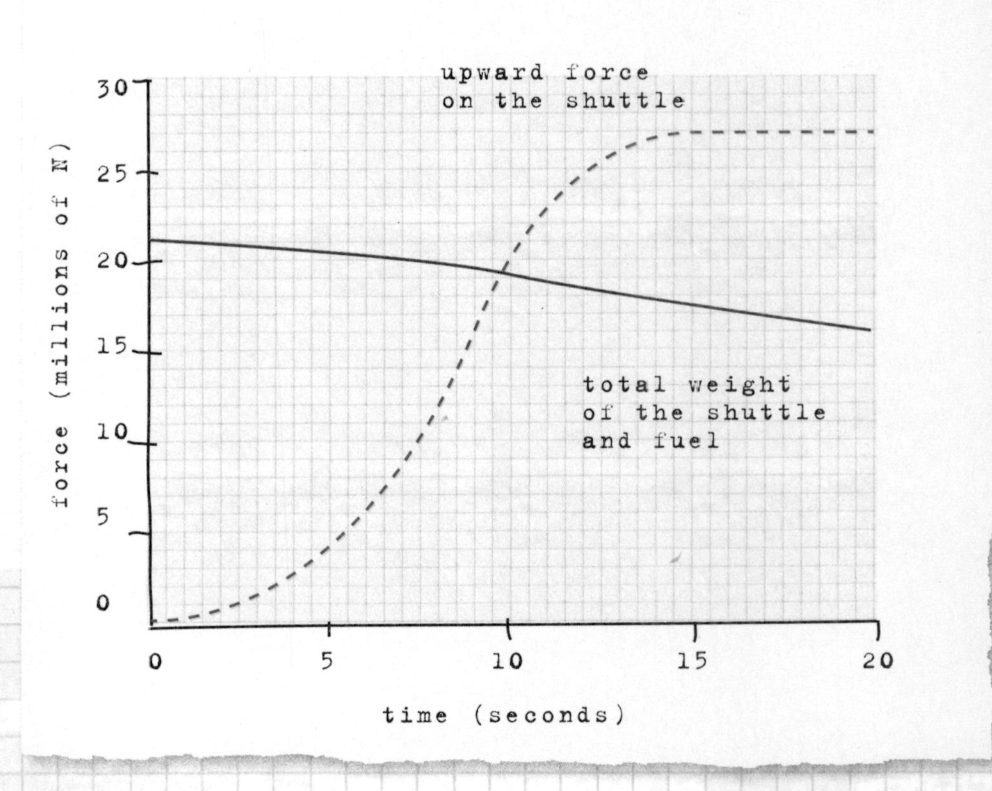

answers p122/3

Label the layers of the Earth.

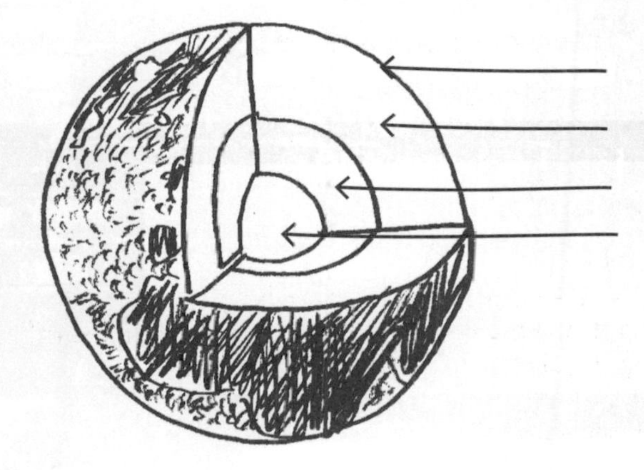

Four gears are connected as shown.

If gear X moves clockwise at 10 rpm, how does gear Y turn?

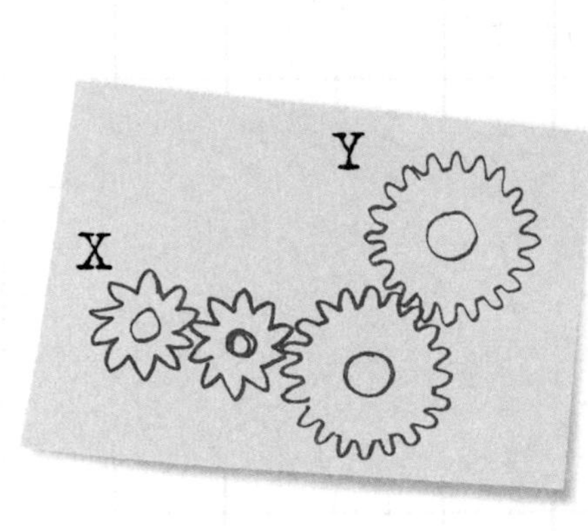

Below are the patterns made by four sound waves on an oscilloscope.

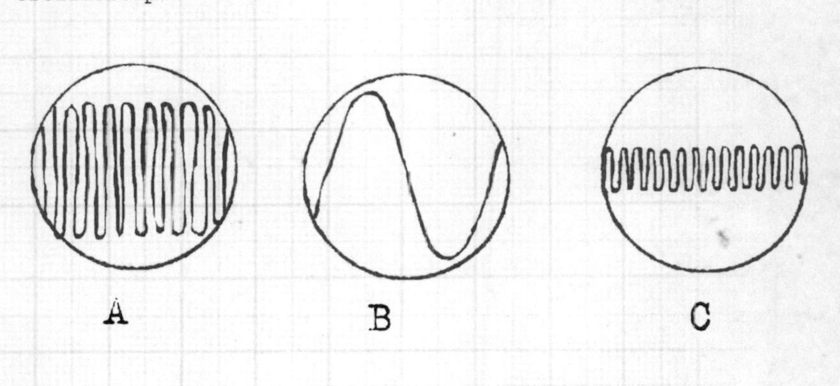

Which pattern shows:

1) A quiet, high-pitched sound?
2) A loud, low-pitched sound?
3) A loud, high-pitched sound?

"My ambition is to live to see all of physics reduced to a formula so elegant and simple that it will fit easily on the front of a T-shirt."

[Leon Lederman]

[answers p122/3]

A car of mass 1,000 kilograms is traveling at a constant speed of 10 m/s. Leaving aside friction, how much force must the engine supply to keep traveling at the same speed?

[answers p122/3]

What is the value of the missing force, F, in the diagram?

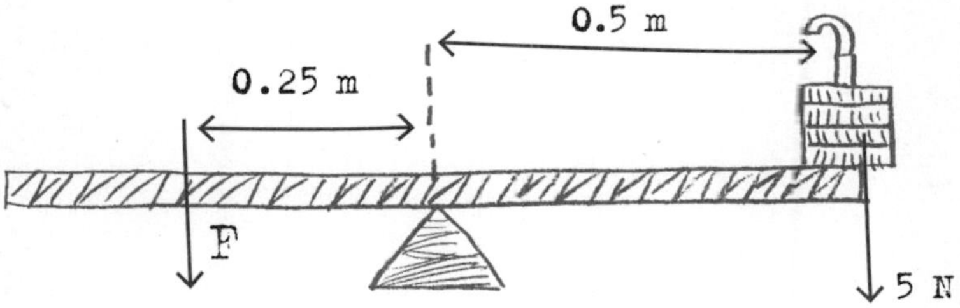

0.5 m

0.25 m

F

5 N

A ray of blue light falls on a glass prism. Draw the path of the blue ray though the prism, and then from the prism to the screen.

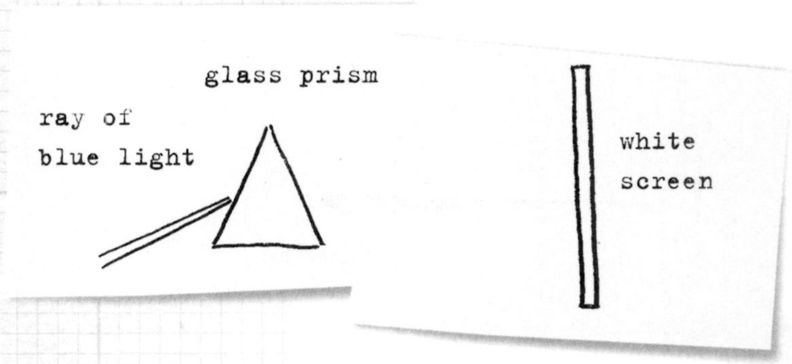

glass prism

ray of
blue light

white
screen

answers p122/3

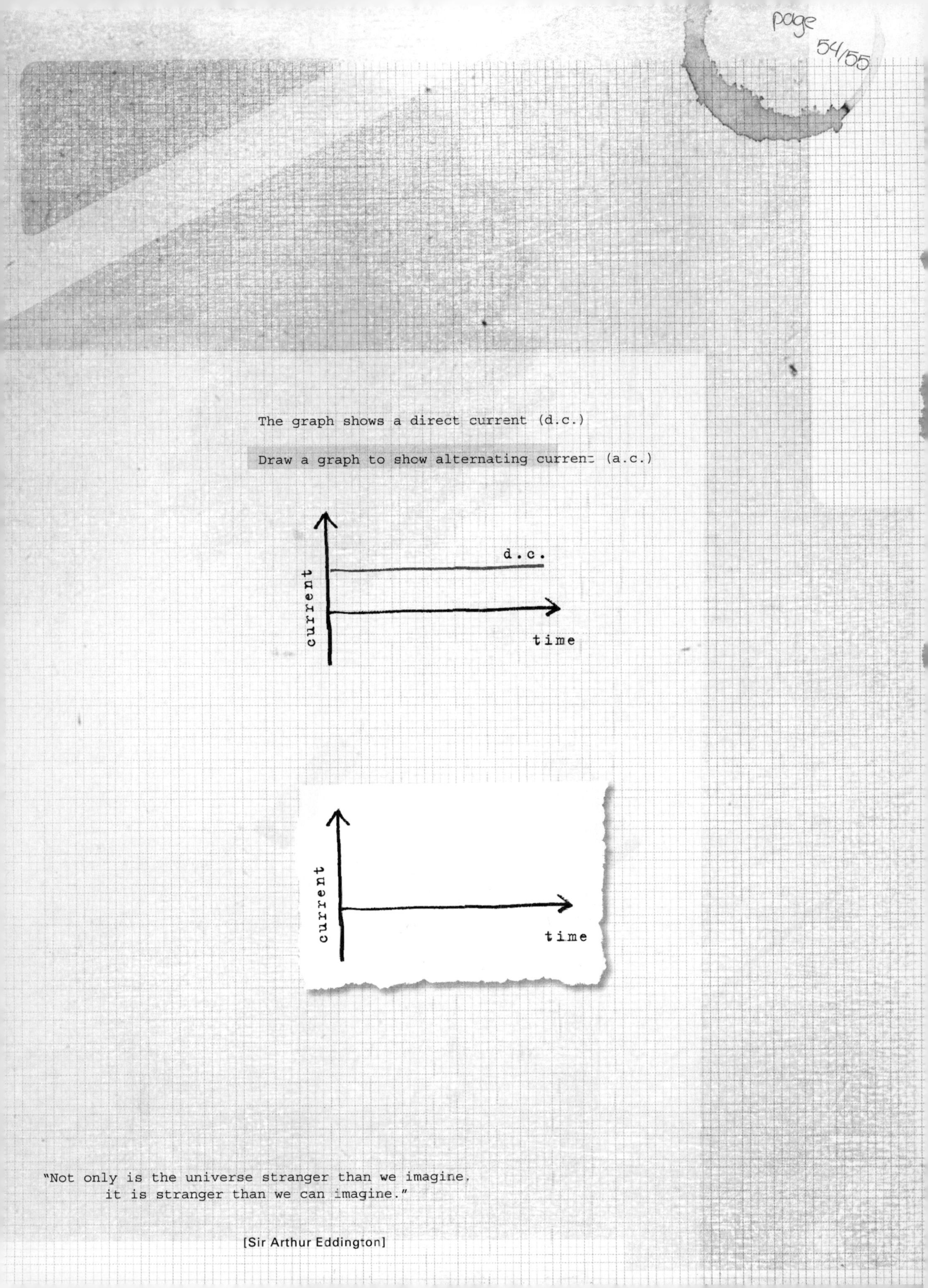
page 54/55
The graph shows a direct current (d.c.)
Draw a graph to show alternating current (a.c.)
d.c.
current
time
current
time
"Not only is the universe stranger than we imagine,
it is stranger than we can imagine."
[Sir Arthur Eddington]

A container is filled with liquid. At each end of the container there is a piston. Piston A has a smaller surface area than piston B. Pushing on the pedal exerts a force of 200 newtons on Piston A.

1) What pressure is exerted on the liquid?
2) If the liquid in the container exerts the same pressure on piston B, what is the force on it?

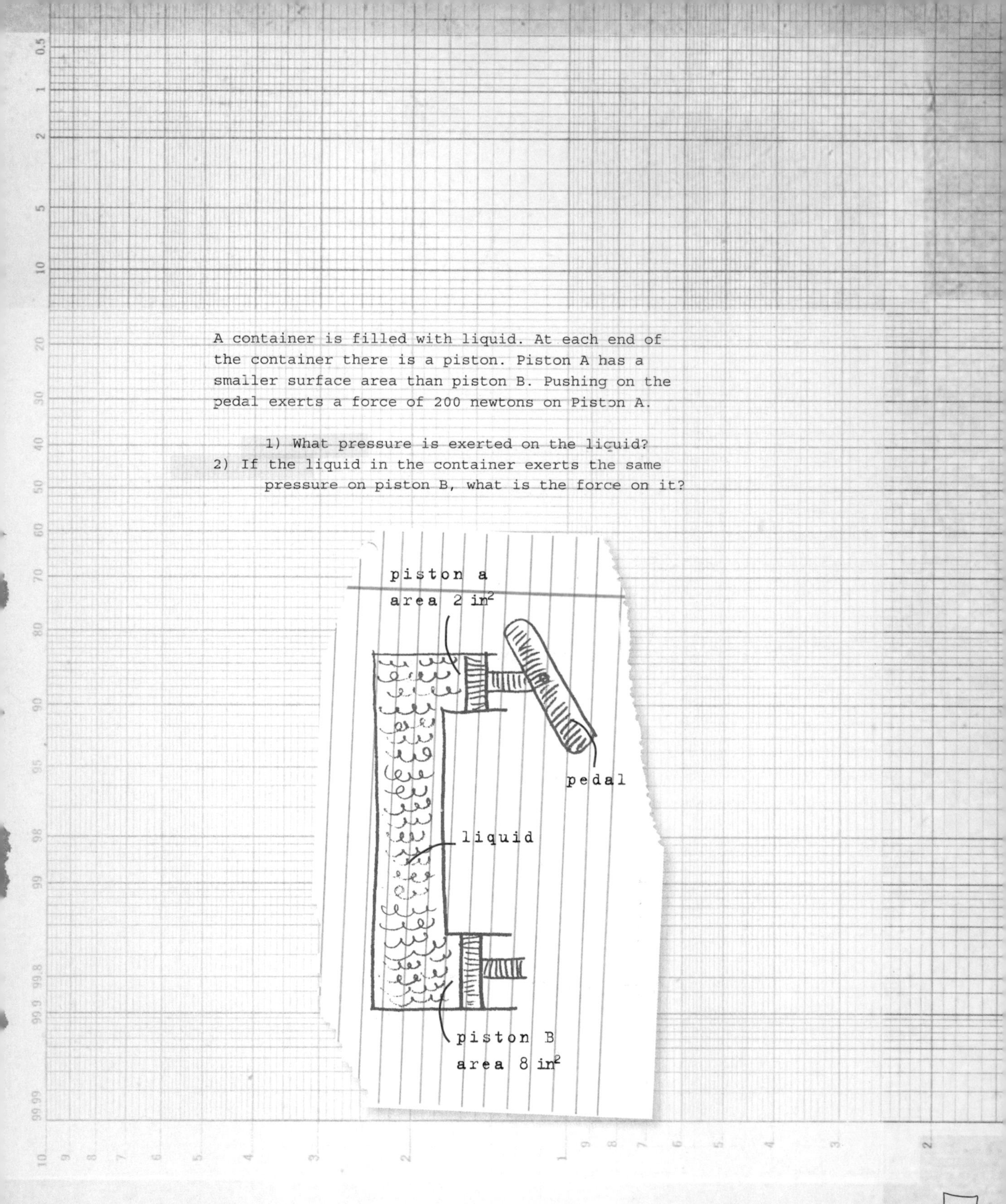

answers p122/3

During a thunderstorm, you see the lightning before you hear the thunder. This is because light travels so fast it is almost instantaneous. Sound, however, travels at 340 m/s in air.

If you hear thunder half a minute after seeing lightning, how far away is the storm?

Here are two weights on a balance.

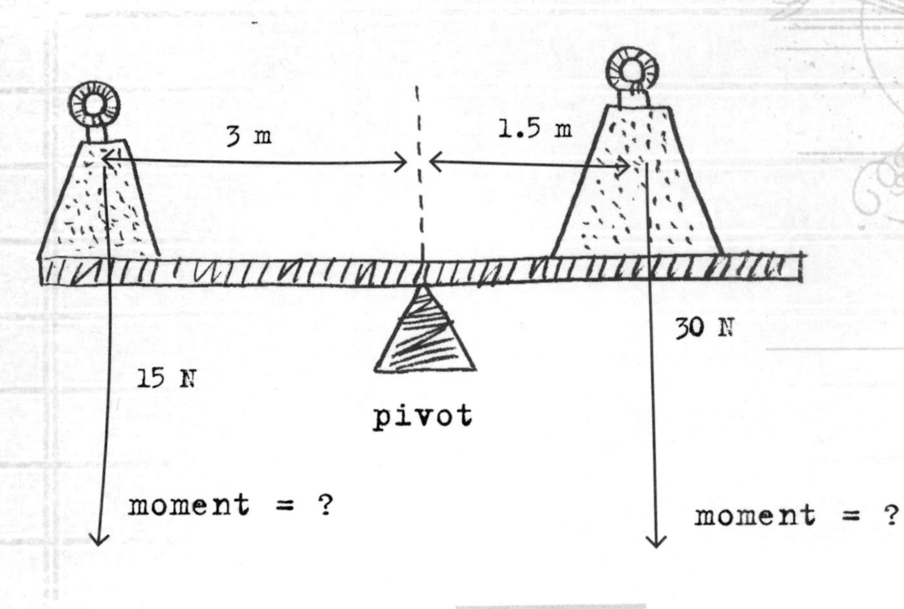

1) What is the moment of weight A?
2) What is the moment of weight B?

"In questions of science, the authority of a thousand is not worth the humble reasoning of a single individual."

[Galileo Galilei]

The ammeter reading in this circuit is 0.5 amps.

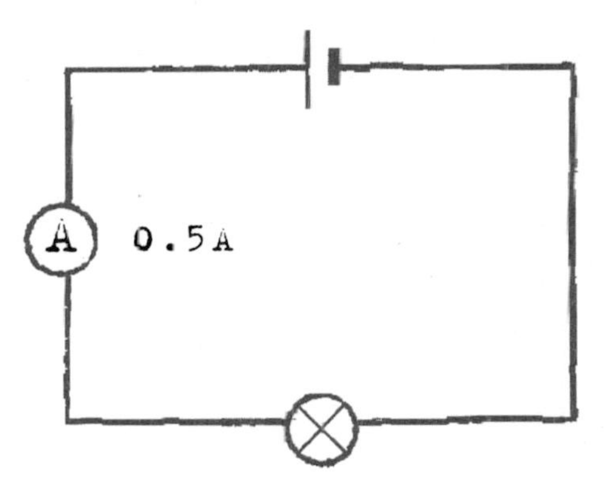

What is the ammeter reading in this circuit?

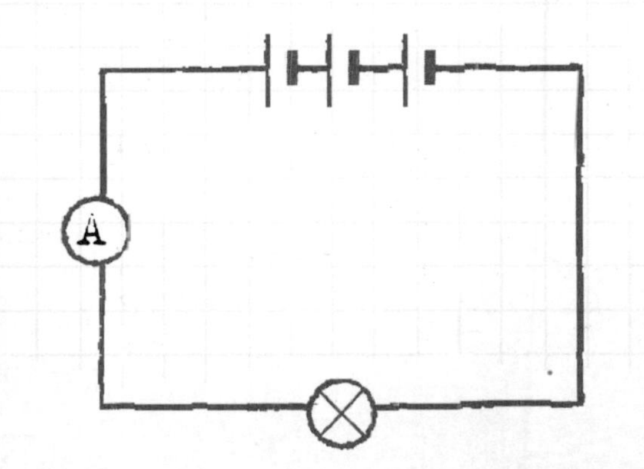

[answers p122/3]

Three machines, X, Y, and Z, are set up to lift a load of 10 lb. E indicates the place where effort is applied to lift the loads. Which machine requires the least effort to lift the load?

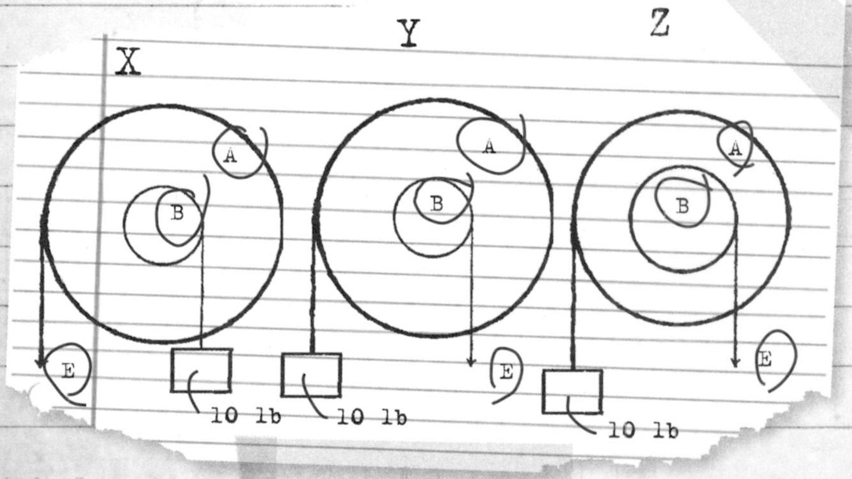

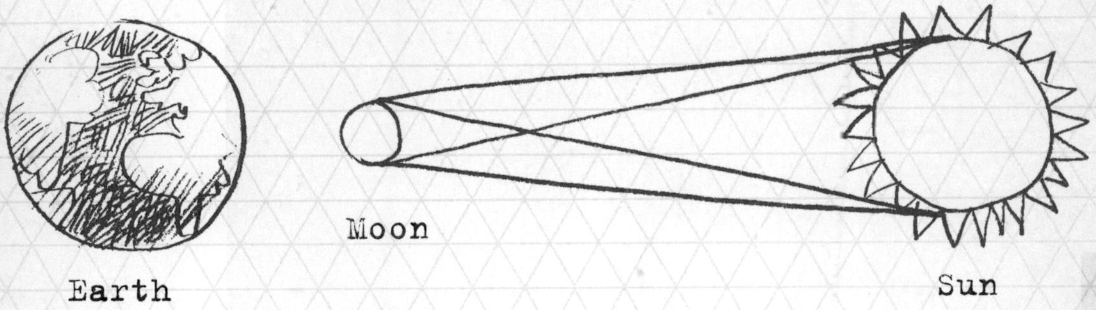

page 60/61

Moon

Earth

Sun

The diagram shows light from the Sun reaching Earth during a solar eclipse.
Continue the rays of light and then shade the area to show where a total eclipse takes place.

answers p122/3

A wave machine in a pool produces 10 waves every second. The waves travel 30 feet along the pool in 4 seconds.

What are the frequency, speed, and wavelength of the waves?

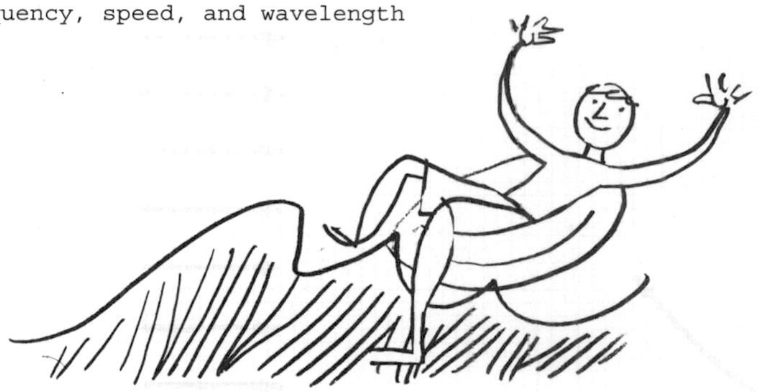

"Everything that can be counted does not necessarily count;
everything that counts cannot necessarily be counted."

[Albert Einstein]

The strength of gravity on the Moon is 1.6 newtons per kilogram.

An astronaut has a mass of 80 kilograms on Earth. What is his mass on the Moon?

answers p122/3

The reading on the first ammeter in this circuit is 0.3 amps.

Label the amp readings on the other three ammeters in the circuit.

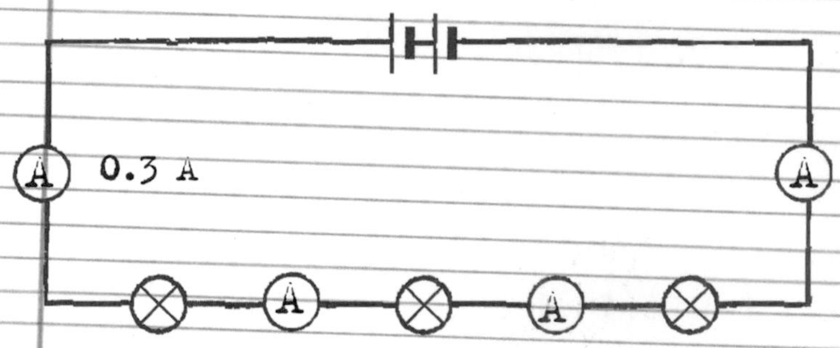

0.3 A

answers p122/3

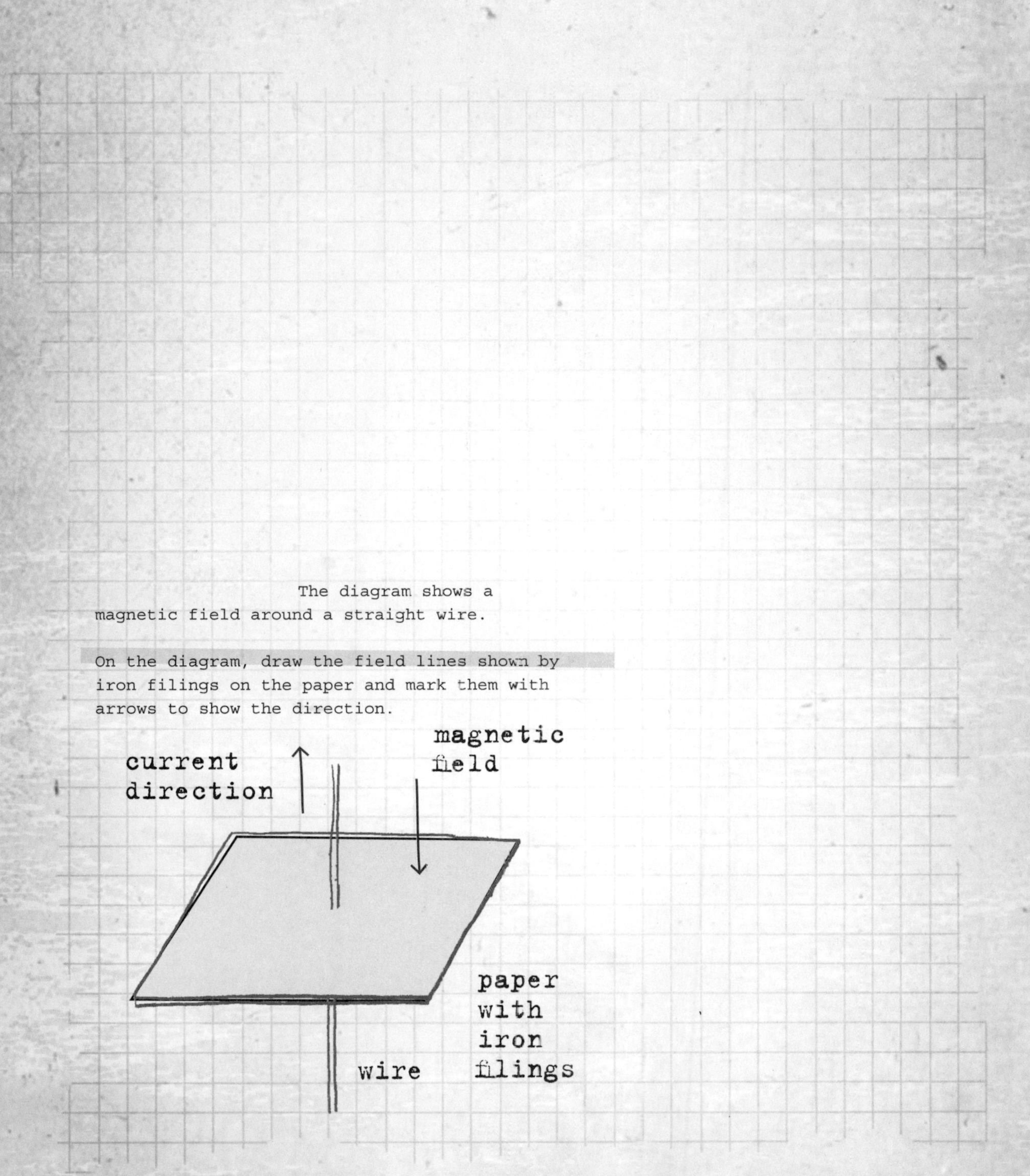
page 64/65
The diagram shows a
magnetic field around a straight wire.
On the diagram, draw the field lines shown by
iron filings on the paper and mark them with
arrows to show the direction.
magnetic
field
current
direction
paper
with
iron
filings
wire

Where do the following waves fit in the electromagnetic spectrum?

infrared   gamma rays   ultraviolet   radio waves

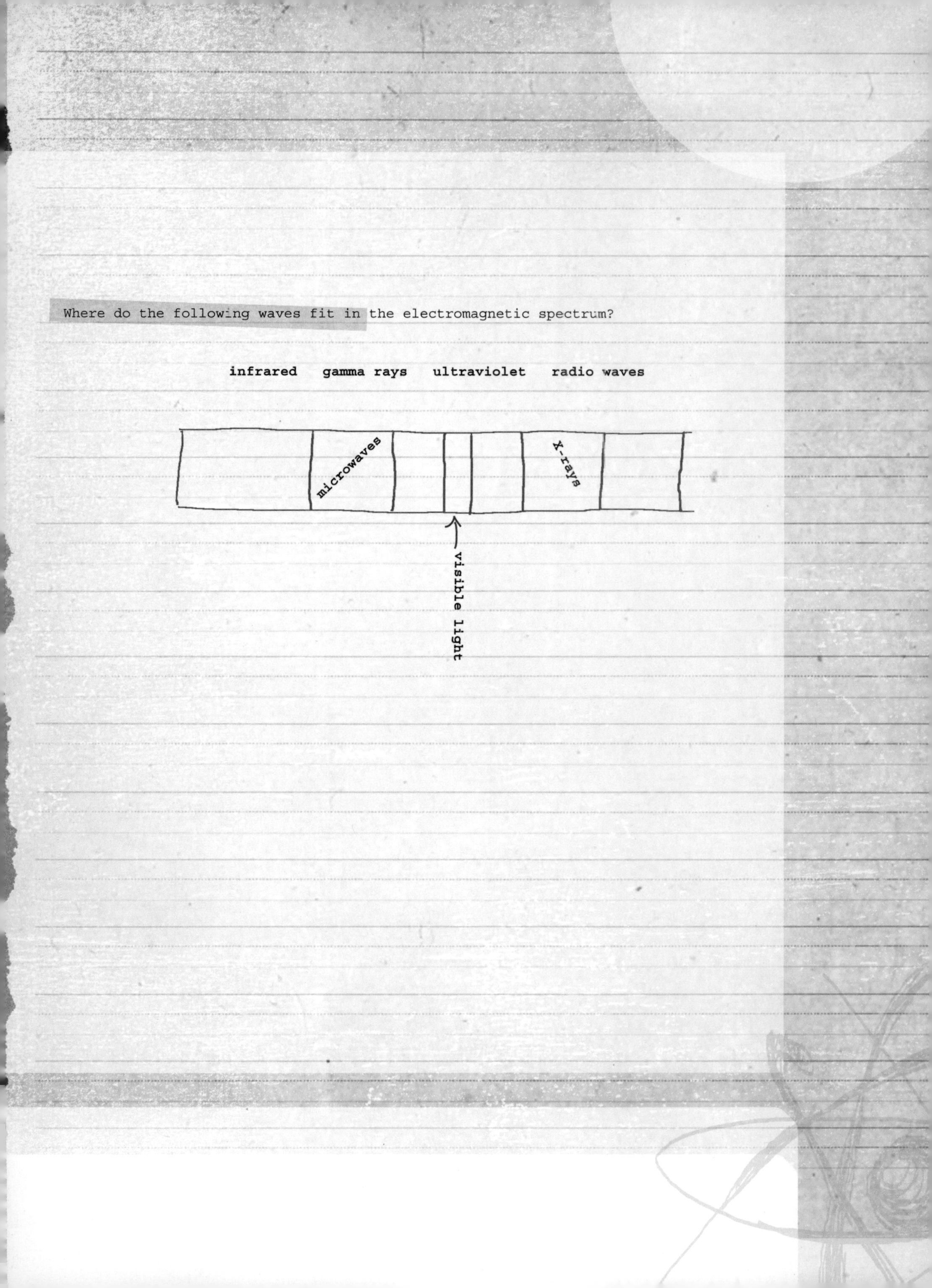

A roller coaster car starts at point A and ends at point F.

1) At which two points does the car have no kinetic energy?
2) At which point does the car have the most gravitational potential energy?
3) What force keeps the car moving between points B and C?

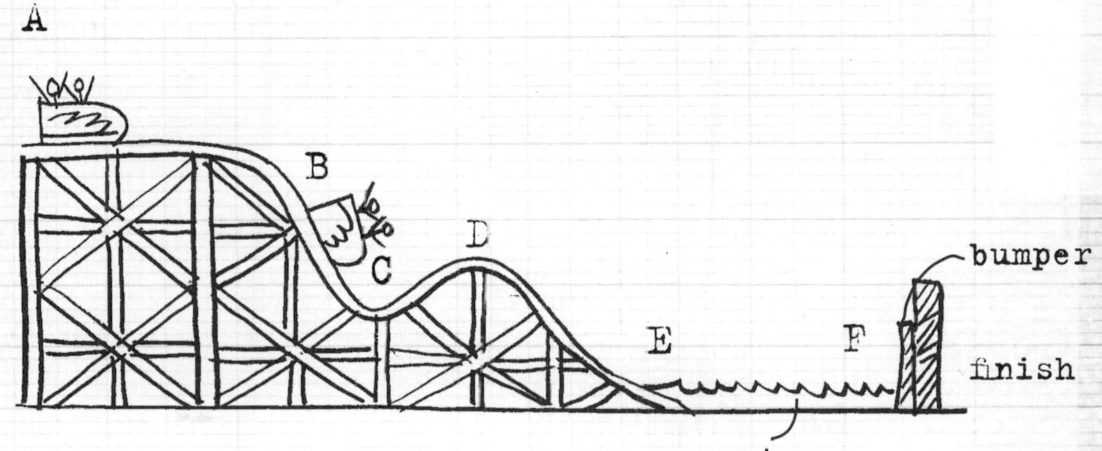

"Gravitation is not responsible for people falling in love."

[Sir Isaac Newton]

answers p122/3

A Galapagos tortoise crawls along a beach at 0.2 miles per hour. How far will it have crawled in 6 hours and 30 minutes?

A ray of light is shone through a glass block.

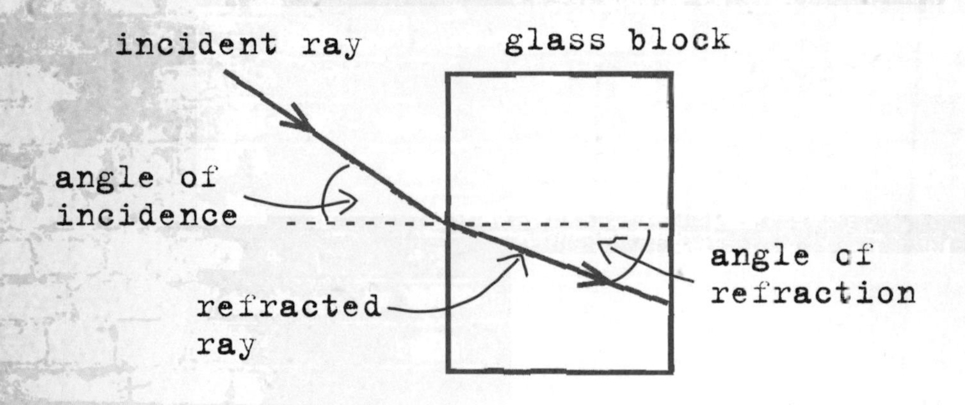

The angles of incidence and the angles of refraction are plotted on a graph.

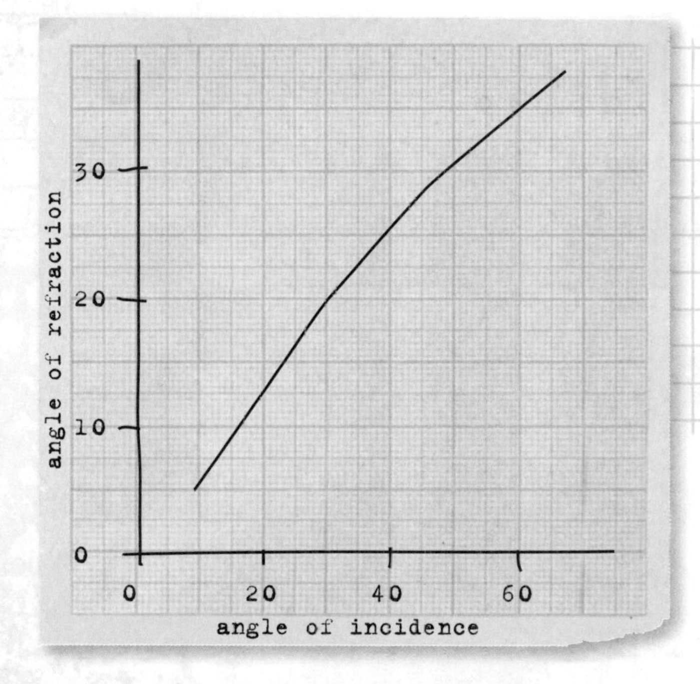

1) What does the graph reveal about the relationship between the angle of incidence and the angle of refraction?

2) When the angle of refraction is 20°, what is the angle of incidence?

[answers p124/5]

At which position on the diagram does the pendulum have maximum kinetic energy?

"Anyone who is not shocked by quantum theory has not understood it."

[Niels Bohr]

At its closest point, the Sun is 91,402,000 miles from Earth. At its farthest point, it is 94,512,000 miles from Earth. Light travels at 186,282 miles per second.

Using the average distance, how long does it take light from the Sun to reach Earth, in minutes and seconds?

answers p124/5

The diagram shows the Doppler effect of sound waves.

Complete the sentences to explain the Doppler shift.

As a source moves toward an observer, the wavelength of the sound waves ______ and their frequency ______

As a sources moves away from an observer, the wavelength of the sound waves ______ and their frequency ______

answers p124/5

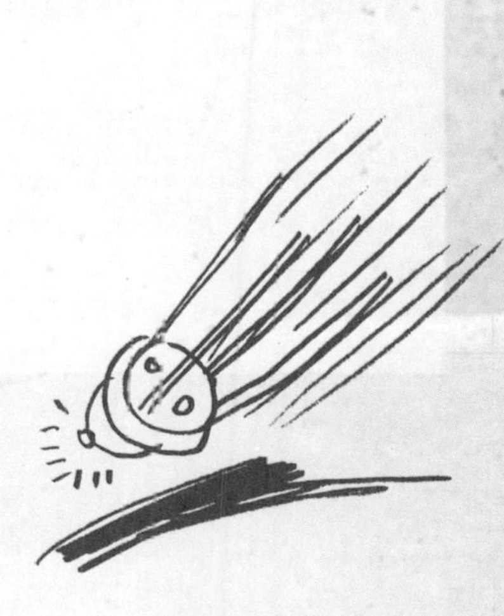

The force of gravity on the surface of Jupiter is around 26 newtons per kilogram. A probe of mass 50 kilograms lands on the surface of the planet.

What is the weight of the probe?

A piece of copper wire is used to connect points C and D in the circuit.

Which of the lamps are lit?

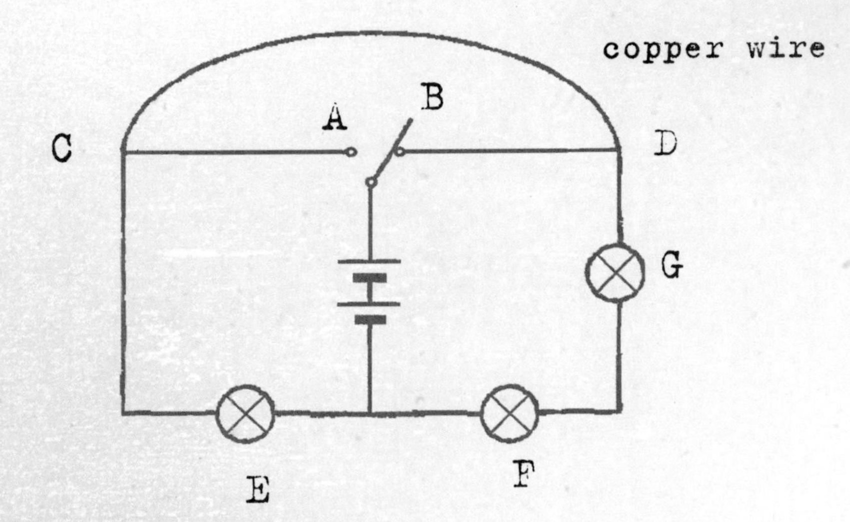

"Something unknown is doing we don't know what."

[Sir Arthur Eddington]

In the following examples, are the forces balanced or unbalanced?

1) A car braking
2) A satellite orbiting Earth at a constant speed
3) A cyclist traveling at a constant speed on a straight road

[answers p124/5]

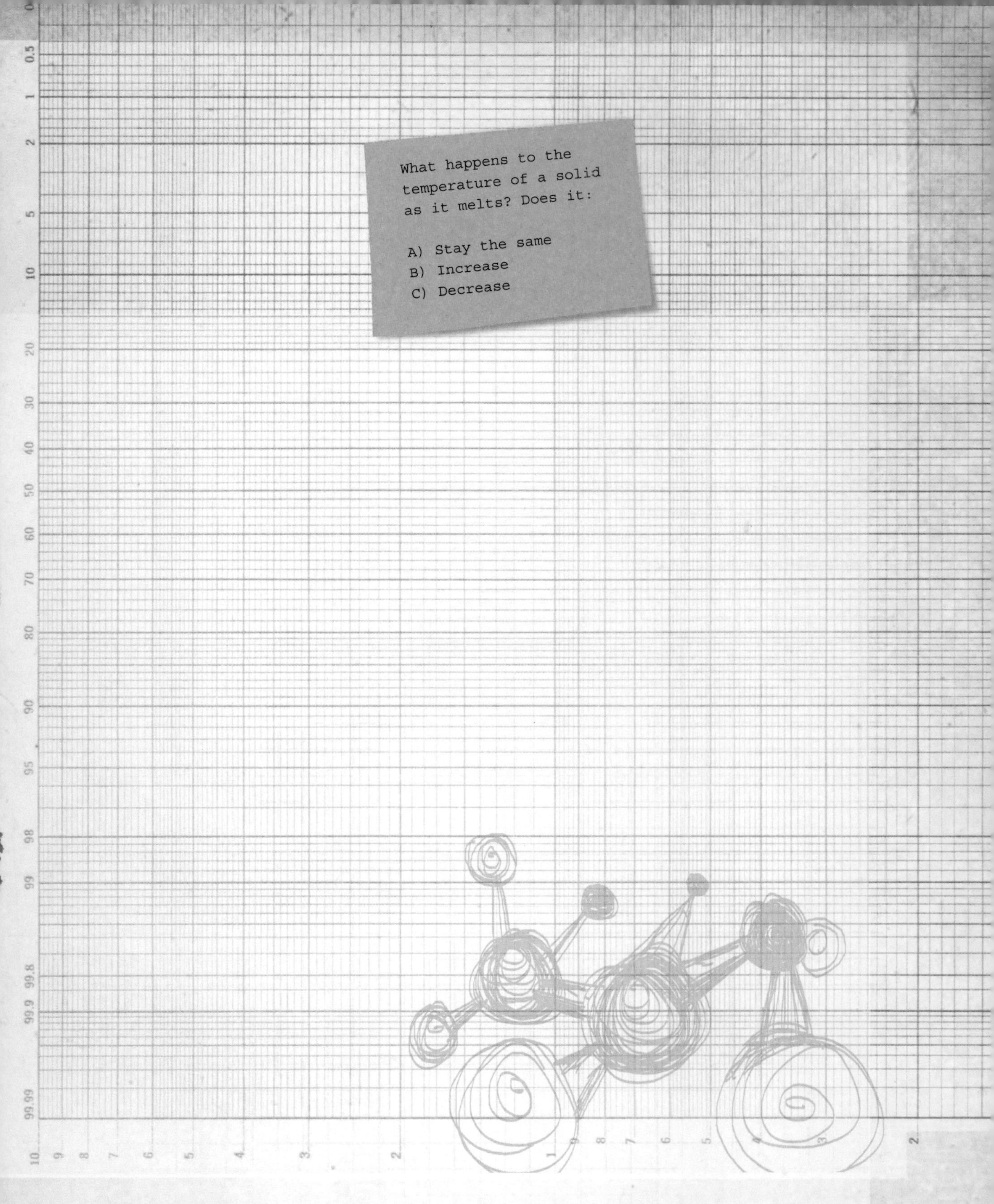

What happens to the temperature of a solid as it melts? Does it:
A) Stay the same
B) Increase
C) Decrease

The table contains information about some of the bodies in the Solar System.

Body	**Average distance from Sun (Astronomical Units)**	**Diameter relative to Earth**
Mercury	0.39	0.38
Earth	1.00	1.00
Jupiter	5.20	11.20
Moon	1.00	0.27

1) How many times larger than the Moon is Earth?
2) How many times farther away from the Sun than Mercury is Jupiter?

answers p124/5

A man is pushing a loaded trolley.
F shows an unbalanced force acting on the trolley.

F

1) What happens to the acceleration if the size of the force F is doubled?
2) What happens to the acceleration if the original force is applied, but the mass of the trolley is doubled?

"An experiment is a question which science poses to Nature and a measurement is the recording of Nature's answer."

[Max Planck]

answers p124/5

The table shows information about the speed of a runner.

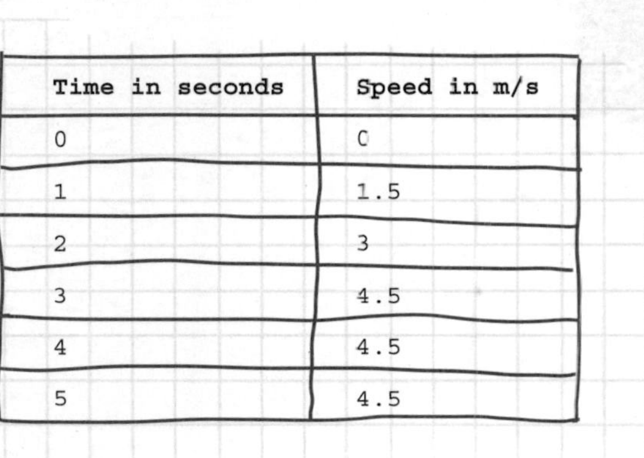

Time in seconds	Speed in m/s
0	0
1	1.5
2	3
3	4.5
4	4.5
5	4.5

1) Plot a graph of this information

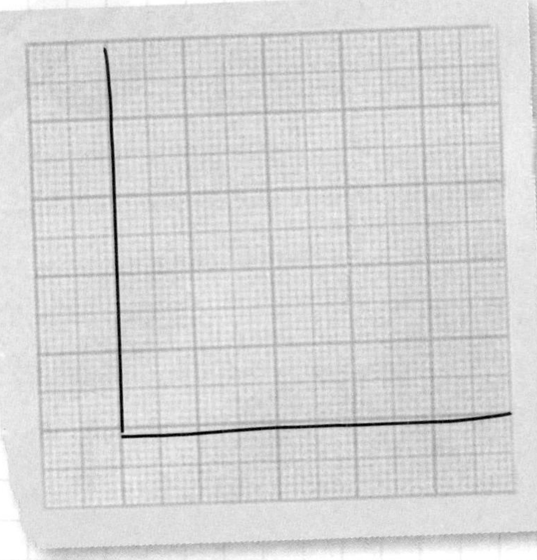

2) What is the runner's acceleration at 3 seconds?

The table shows the power rating of some domestic appliances.

Appliance	**Power rating (watts)**	**Average daily use**
Kettle	2,000	10 minutes
Iron	1,200	30 minutes
Vacuum cleaner	1,000	20 minutes

Complete the table to show the energy used by each appliance in joules and kilowatt-hours.

Appliance	**Energy in joules**	**Energy in kWh**
Kettle		
Iron		
Vacuum cleaner		

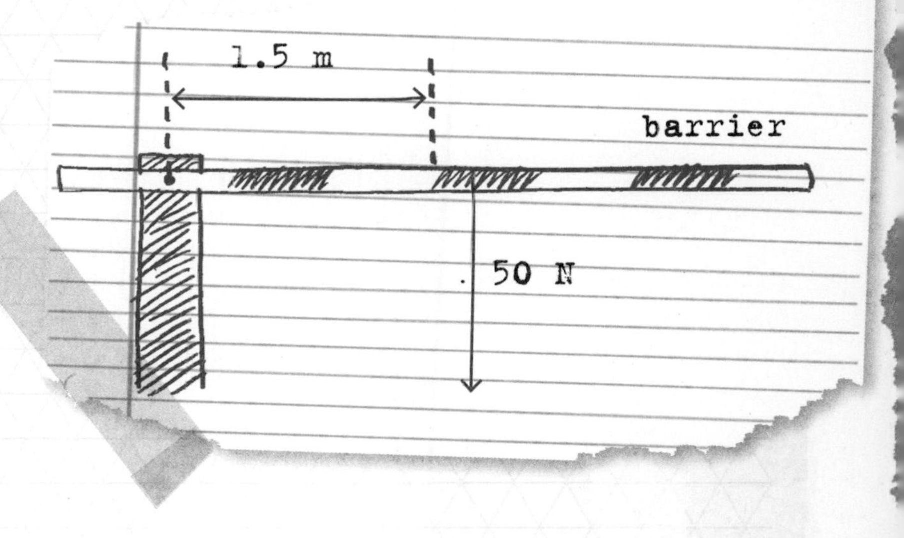

The diagram shows a barrier in a parking lot.

What is the turning moment of the barrier about the pivot?

answers p124/5

It costs $7,000 to install double glazing in a house. The annual saving on energy bills as a result of this is $350. What is the payback time?

A ship sends out ultrasound pulses to calculate the depth of the sea at a particular point. The speed of the ultrasound pulse is 5,000 feet per second. The time taken for a pulse to return is 20 milliseconds.

Calculate the depth of the water.

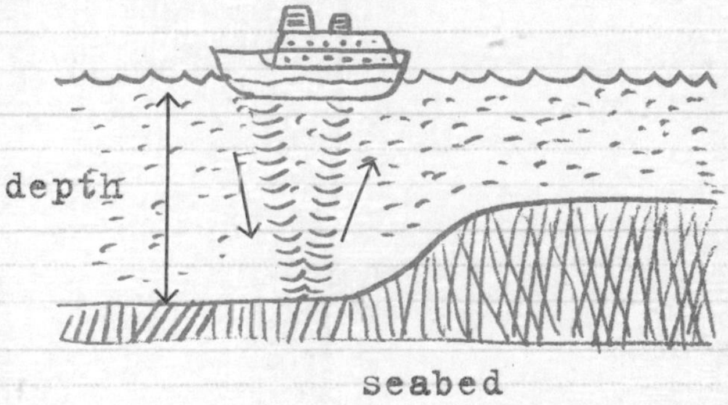

depth

seabed

answers p124/5

"Every day sees humanity more victorious in the struggle with space and time."

[Guglielmo Marconi]

An archer fires an arrow. After 0.5 seconds, the arrow reaches a speed of 150 m/s.

What is the acceleration of the arrow?

This is a Sankey diagram for an energy-saving lightbulb.

What type of energy does the downward arrow represent, and what is its value?

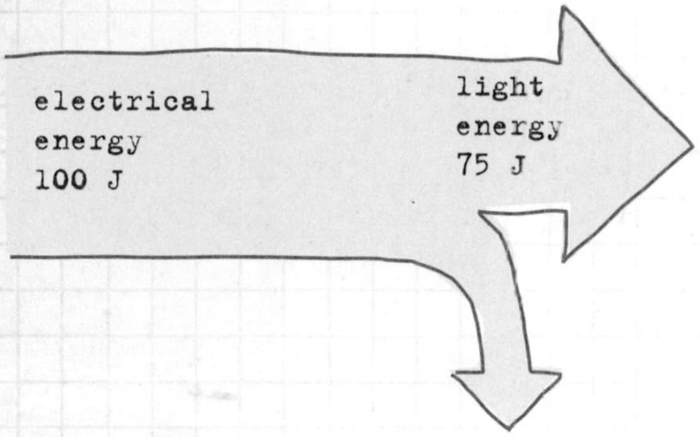

electrical
energy
100 J

light
energy
75 J

answers p124/5

A flashlight lamp has a resistance of 3.5 ohms. What is the potential difference across the flashlight lamp when a current of 2.0 amps flows through it?

"If all of mathematics disappeared, physics would be set back by exactly one week."

[Richard Feynman]

Complete the table to show the type of energy produced by each device and whether each gives digital or analog output.

Device	Energy produced	Digital or analog
Loudspeaker		
Electric motor		
Light-emitting diode (LED)		

answers p124/5

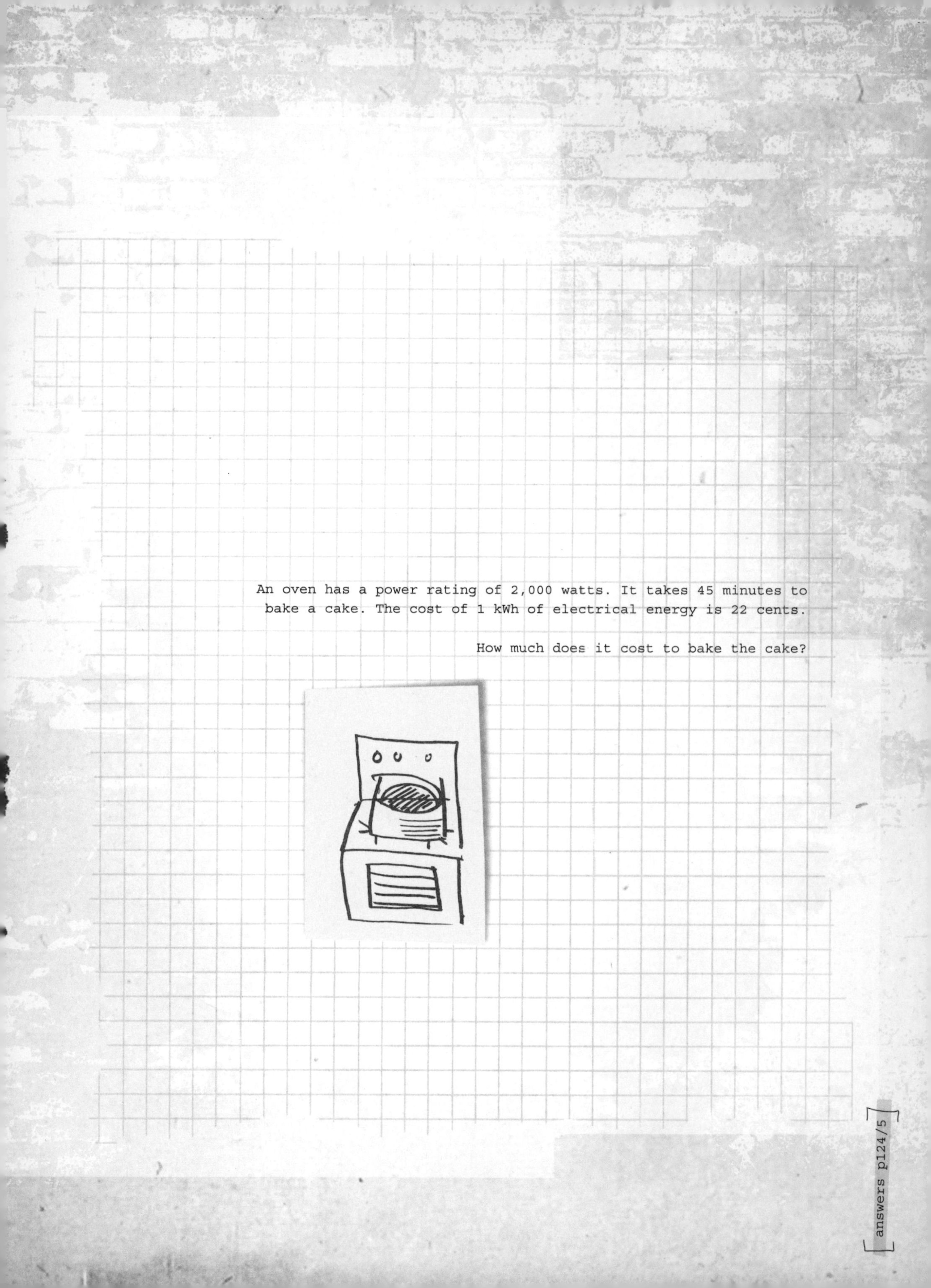

An oven has a power rating of 2,000 watts. It takes 45 minutes to bake a cake. The cost of 1 kWh of electrical energy is 22 cents.

How much does it cost to bake the cake?

answers p124/5

The oscilloscope screen shows the pattern produced when a clarinet is played near a microphone.

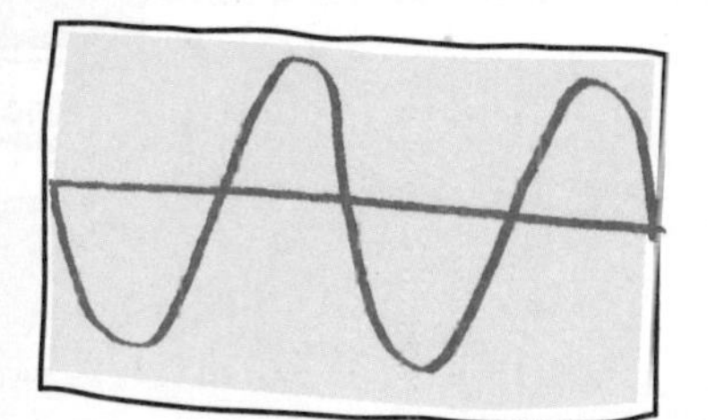

On the oscilloscope screen below, draw the pattern that would appear if the clarinet was played at a lower volume but a higher pitch.

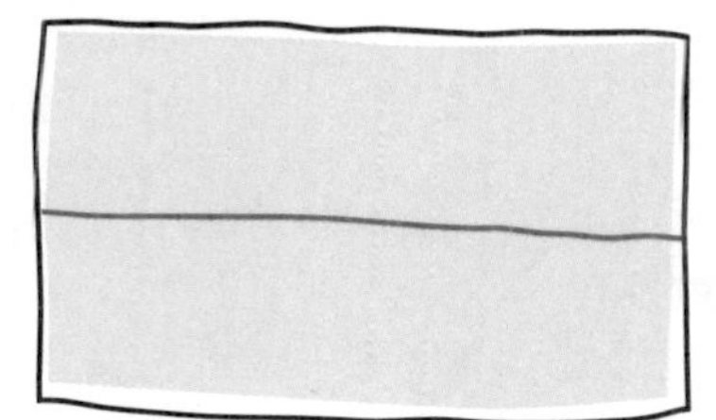

A child of mass 55 kilograms is on a skateboard. He experiences a force of 80 newtons. What will be the child's acceleration?

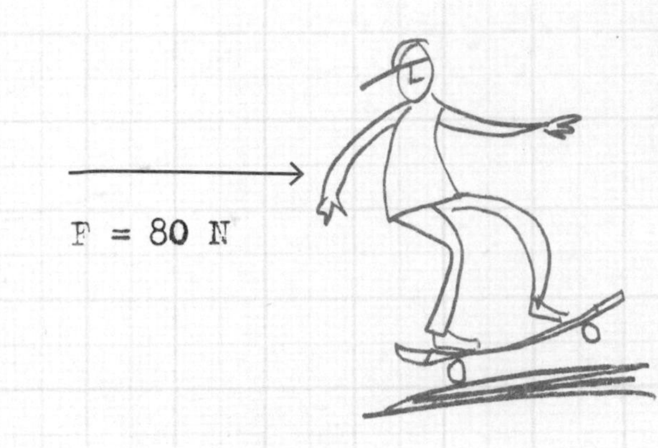

Using the information in the table, mark the positions where each planet will be six months after those shown on the diagram.

Planet	Time to orbit the Sun in Earth years	Distance from the Sun in million km
Mercury	0.25	60
Venus	0.5	108
Earth	1.0	150
Mars	2.0	228

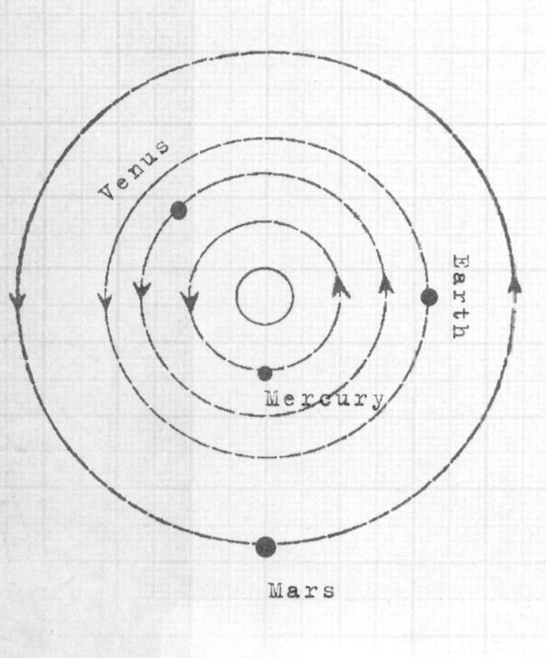

answers p124/5

"There's two possible outcomes: if the result confirms the hypothesis, then you've made a discovery. If the result is contrary to the hypothesis, then you've made a discovery."

[Enrico Fermi]

Electrical current is a movement of charge through a material. Charge is measured in coulombs.

The current passing through a lamp is 1.5 amps. How much charge passes through the lamp in three minutes?

[answers p124/5]

This is a velocity-time graph for two cars.

1) Which car has the greatest acceleration, and how do you know?
2) Which car has traveled farthest, and how do you know?

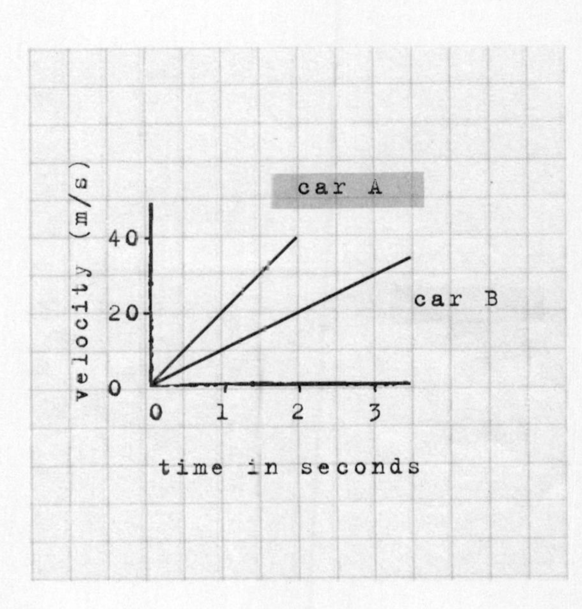

In steel, sound waves travel at 2 km/s.

What is the wavelength of a sound wave in steel that has a frequency of 80,000 Hz?

There are three main types of radiation: alpha, beta, and gamma.

1) Which type is an electromagnetic wave?
2) Which type is a helium nucleus?
3) Which type is an electron?

"Scientific discovery and scientific knowledge have been achieved only by those who have gone in pursuit of it without any practical purpose whatsoever in view."

[Max Planck]

answers p124/5

The graph shows the thinking (A) and braking (B) distances of a car with an initial velocity of 20 m/s.

If the same car was traveling at 30 m/s, what would the thinking distance be?

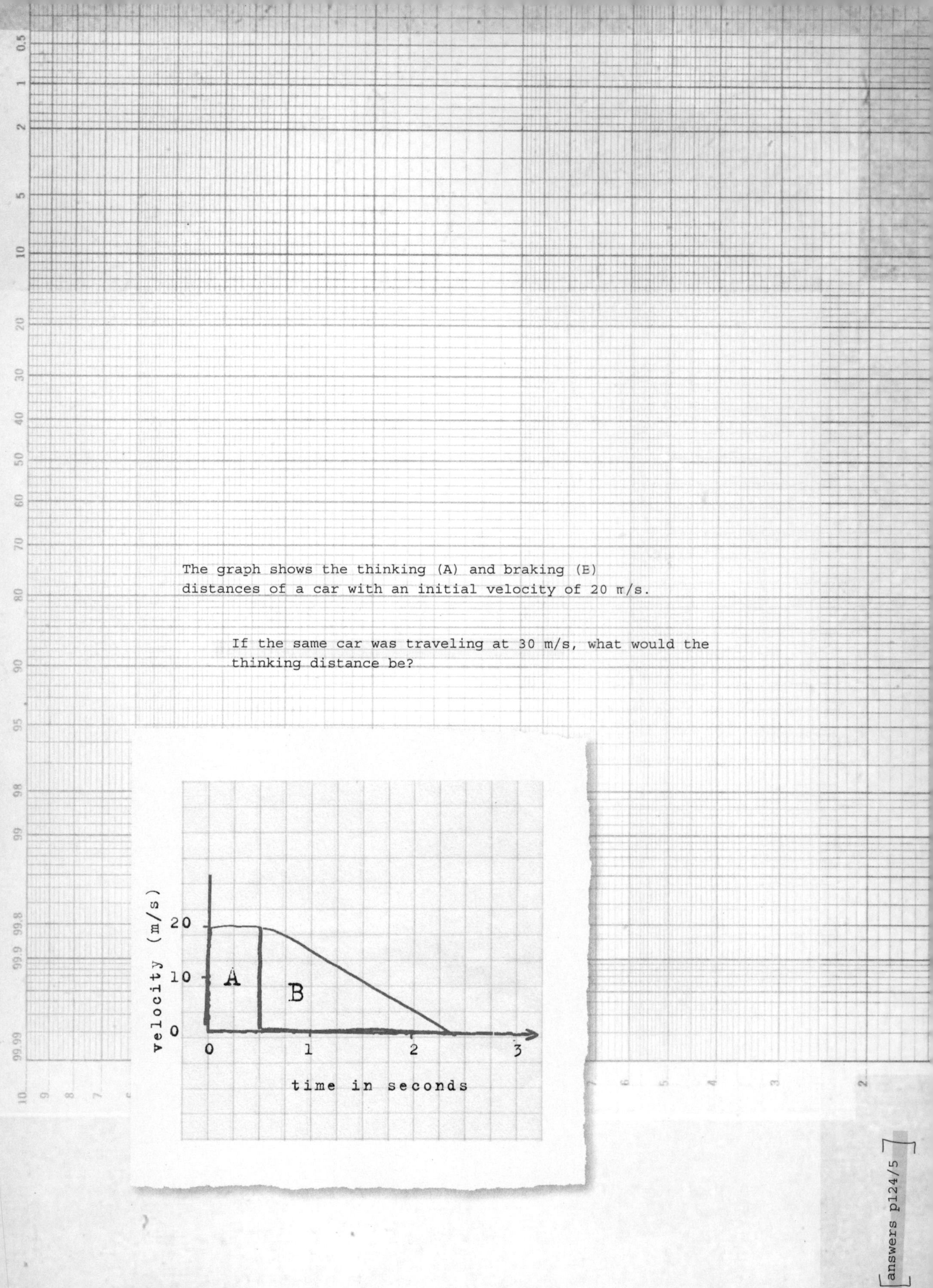

answers p124/5

Here are some water waves. They are approaching a barrier with a gap in it. Doodle the waves as they come out the other side of the barrier.

"Anyone who expects a source of power from the transformation of the atom is talking moonshine."

[Ernest Rutherford]

When you rub a balloon on your sweater and then place it next to the wall, the balloon will stick to the wall.

1) Which particle are transferred in the process?
2) What charge does the balloon have after being rubbed on your sweater?

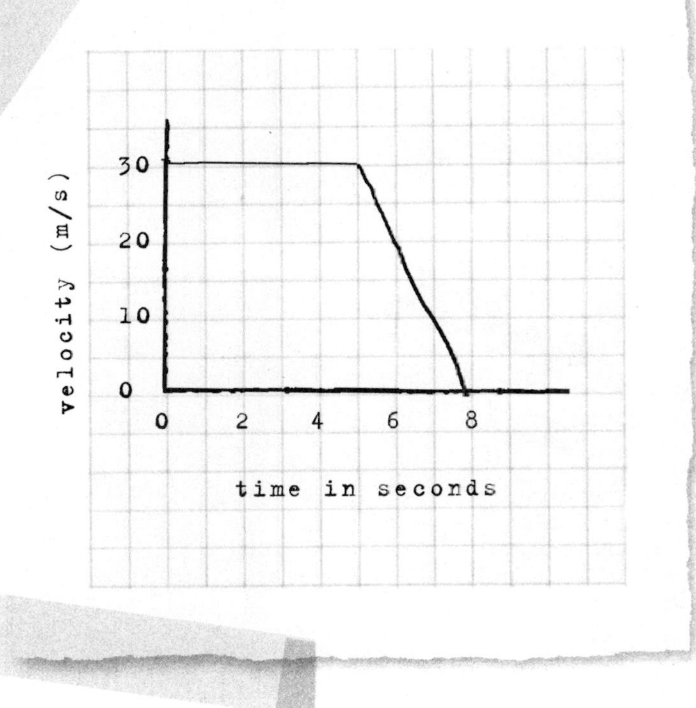

The graph shows how the velocity of a motorbike changes over time.

1) How far does the motorbike travel in the first 5 seconds?
2) What is the deceleration of the motorbike?

answers p125/6

The table shows information about the atomic structure of four substances.

Substance	Protons	Neutrons	Electrons
A	84	124	84
B	85	126	85
C	84	126	84
D	83	126	83

Which two substances are isotopes of the same element?

The graph plots the current and voltage of a filament lamp.

What is the resistance when the voltage is 3?

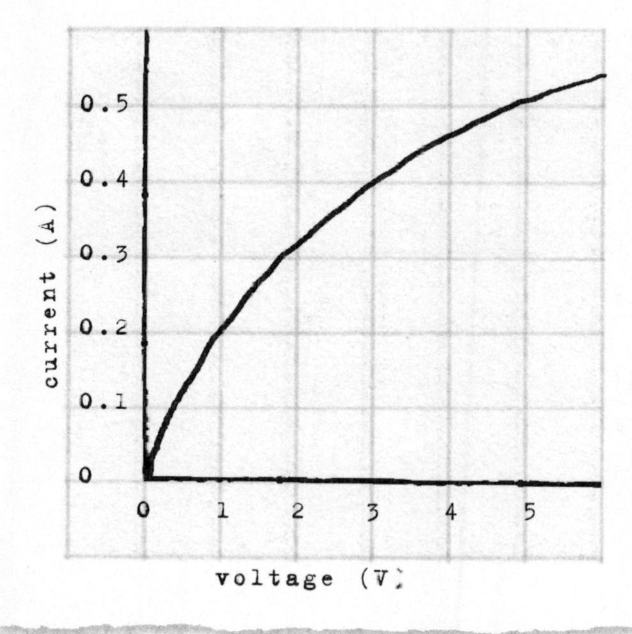

answers p126/7

Which of these diagrams shows a digital signal?

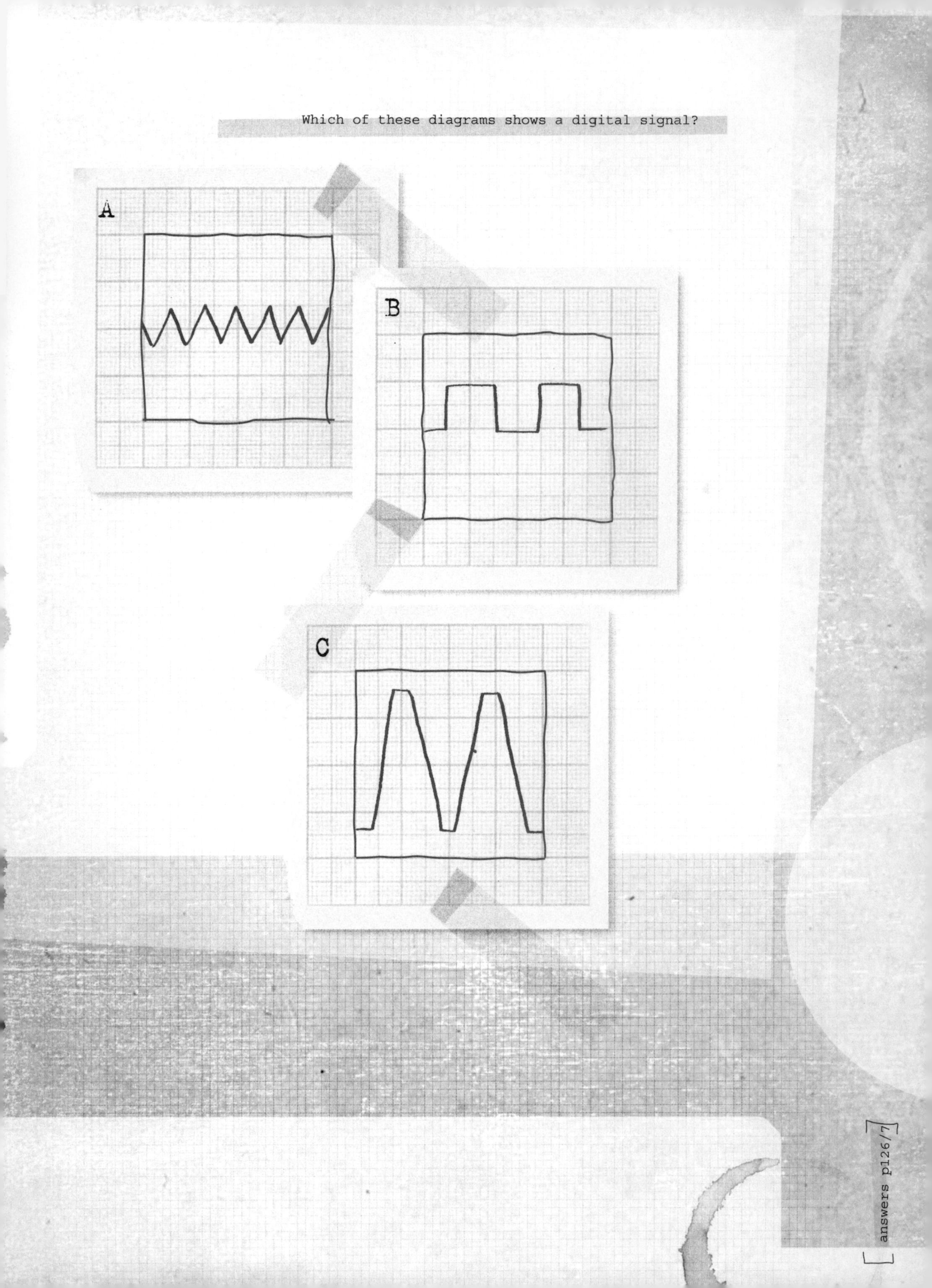

answers p126/7

page 102/103

What is the kinetic energy of a remote-controlled car of mass 12 kilograms traveling at 2.5 m/s

"Nothing is too wonderful to be true if it be consistent with the laws of nature."

[Michael Faraday]

Electromagnetic waves travel at a speed of 3×10^8 m/s in a vacuum. If the frequency of an EM wave is 60 MHz, what is the wavelength of the wave?

Have a bonus point if you can name the type of electromagnetic wave this is!

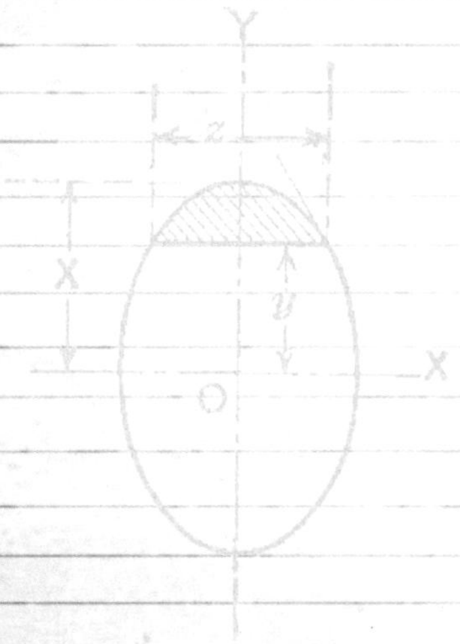

An experiment is carried out to measure the speed of sound. A starting pistol is fired.

Five people stand together 200 meters away and measure the time between seeing the smoke and hearing the shot.

The results they recorded were: 0.55 seconds, 0.62 seconds, 0.58 seconds, 0.56 seconds, and 0.58 seconds.

Work out the average time, then calculate the speed of sound.

answers p126/7

An electric kettle is connected to the mains, which has a voltage of 230. If a current of 10 amps flows through the kettle, what is its power output in kilowatts?

"So far as I know, no one has yet pointed out that the distance travelled in equal intervals of time, by a body falling from rest, stand to one another in the same ratio as the odd number beginning with 1."

[Galileo Galilei]

[answers p126/7]

Here are the forces acting on a ship.

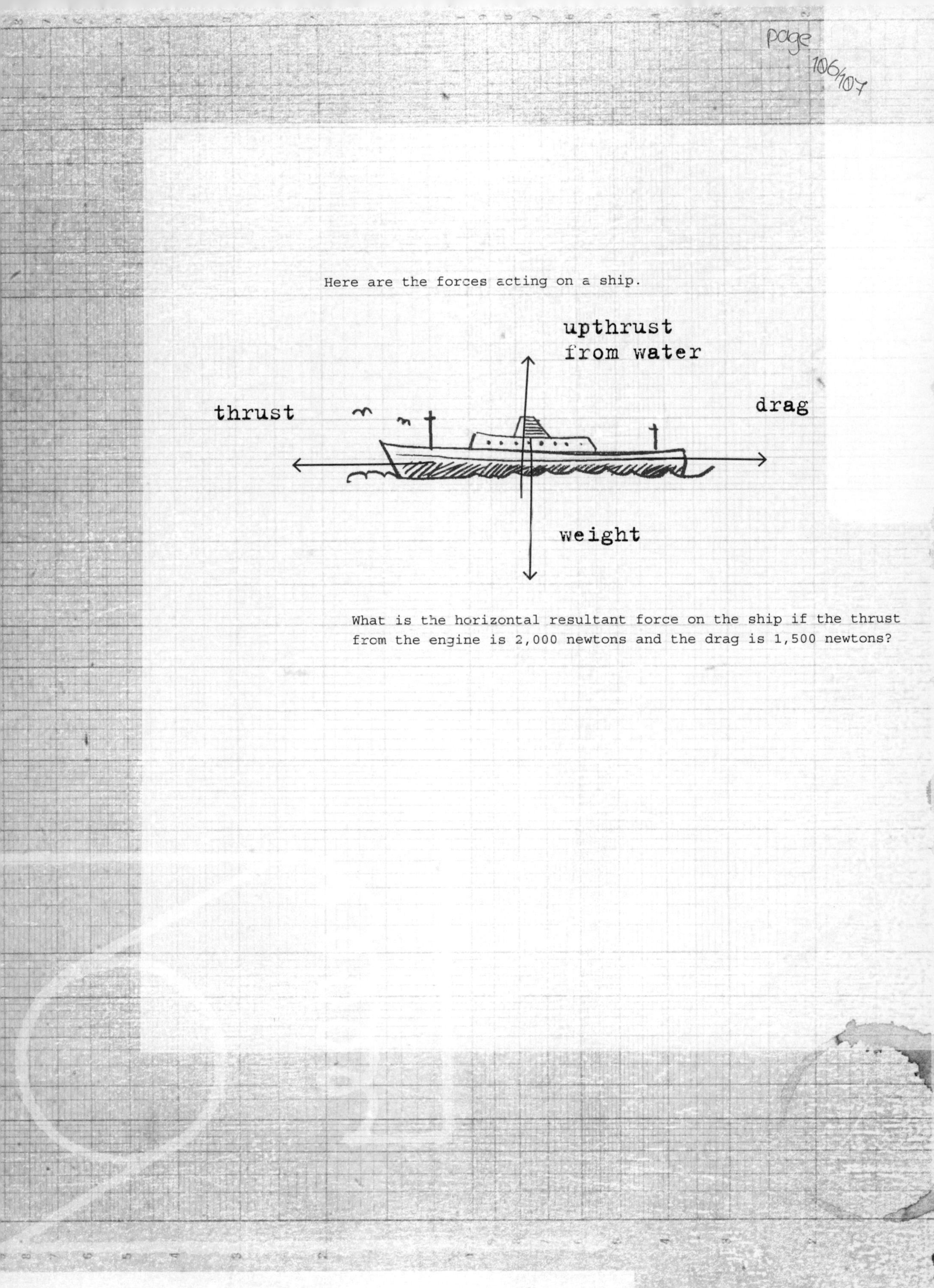

What is the horizontal resultant force on the ship if the thrust from the engine is 2,000 newtons and the drag is 1,500 newtons?

You set out for a walk. You travel 50 meters south, 30 meters west, then 50 meters north. It takes you 50 seconds.

1) What is your final displacement
2) What is your average speed?

page 108/109

What is the lowest frequency in the normal range of human hearing?

A) 20 Hz
B) 200 Hz
C) 20,000 Hz

answers p126/7

An energy-saving lamp has an efficiency of 18%. The output light energy is 10 joules. What is the input electrical energy, calculated to one decimal place?

"It is the weight, not numbers of experiments that is to be regarded."

[Sir Isaac Newton]

page 110/111

The graph shows the velocity against time for a ball thrown vertically upwards.

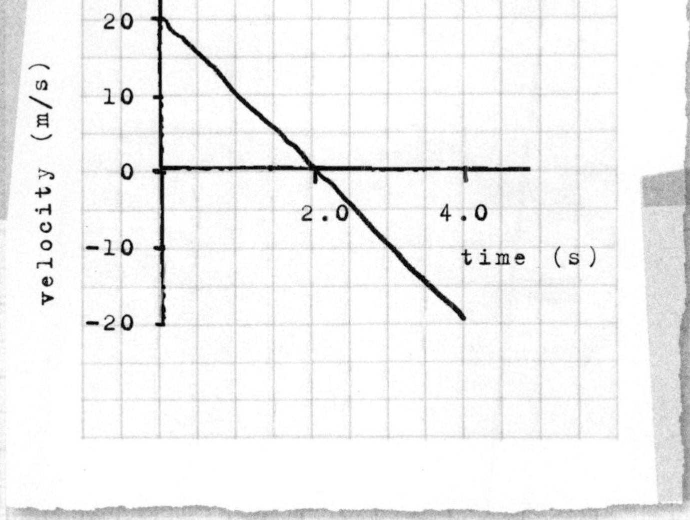

1) After how many seconds does it reach its maximum height?
2) What is the maximum height that ball reaches?

answers p126/7

The filament of a bicycle lamp has a resistance of 3 ohms. It takes a current of 0.6 amps.

What voltage does it work at?

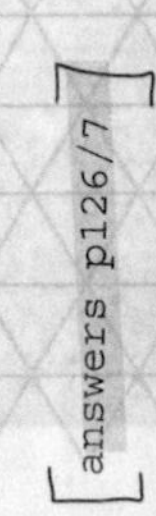

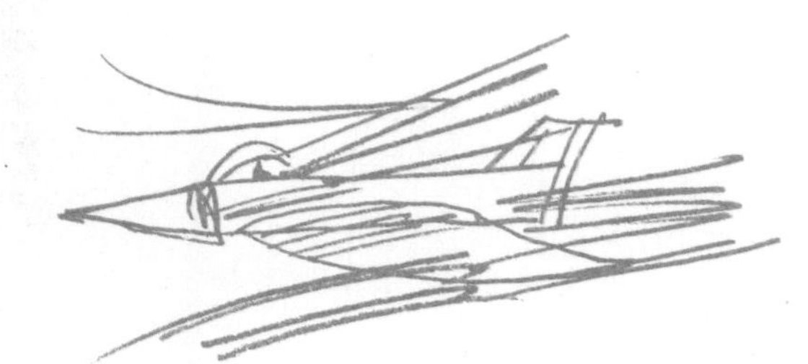

An airplane on a test flight travels at a speed of mach 2 (twice the speed of sound). It travels 1.02 kilometers in 1.5 seconds.

1) What is the average speed of the airplane in m/s?
2) What is the speed of sound?

"I haven't failed, I've found 10,000 ways that don't work."

[Thomas Edison]

"We are just an advanced breed of monkeys on a minor planet of a very average star. But we can understand the Universe. That makes us something very special."

[Stephen Hawking]

A car traveling at 30 m/s has a braking distance of 75 meters.

If the mass of the car is 1,000 kilograms, what braking force was applied?

[answers p126/7]

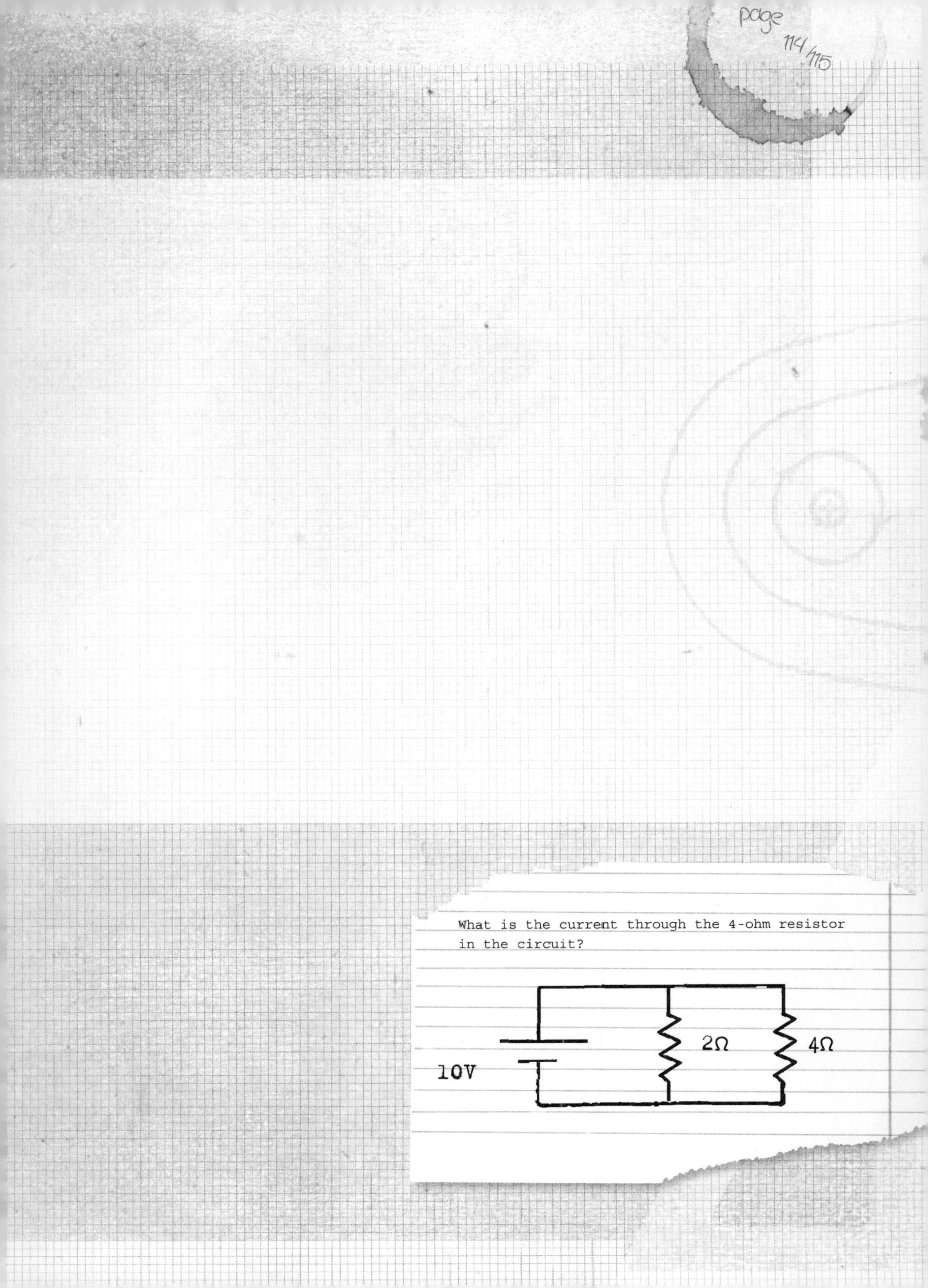
page
114/115
What is the current through the 4-ohm resistor
in the circuit?
10V
2Ω
4Ω

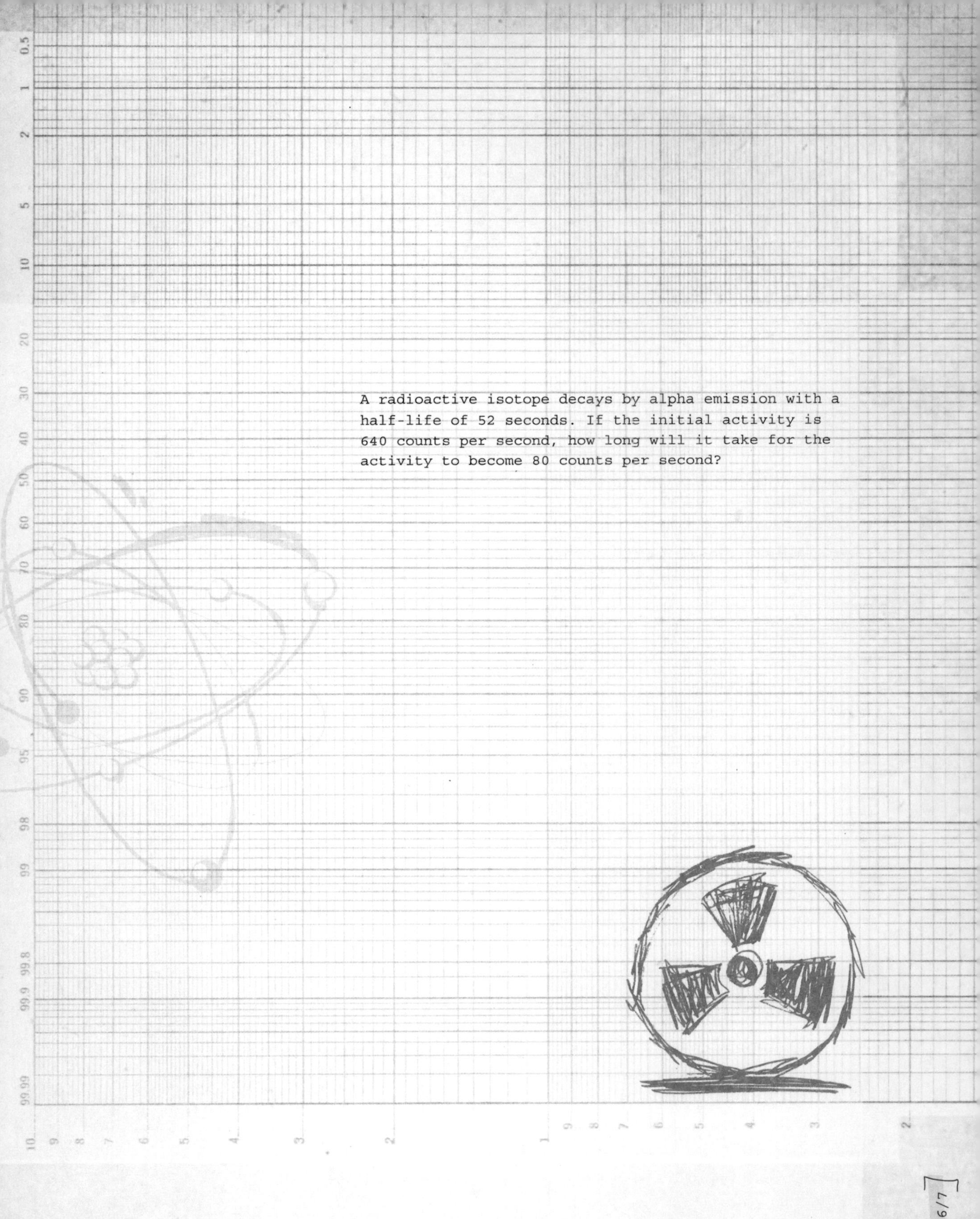

A radioactive isotope decays by alpha emission with a half-life of 52 seconds. If the initial activity is 640 counts per second, how long will it take for the activity to become 80 counts per second?

[answers p126/7]

"Both the man of science and the man of action live always at the edge of mystery, surrounded by it."

[J. Robert Oppenheimer]

A car of mass 800 kilograms is traveling at 25 m/s.

What force is required to stop the car in 50 meters?

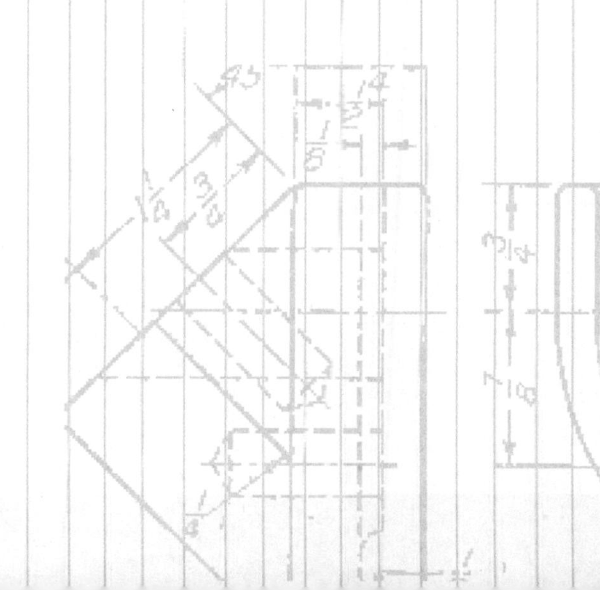

Continue the path of the light ray as it passes through these two right-angled prisms.

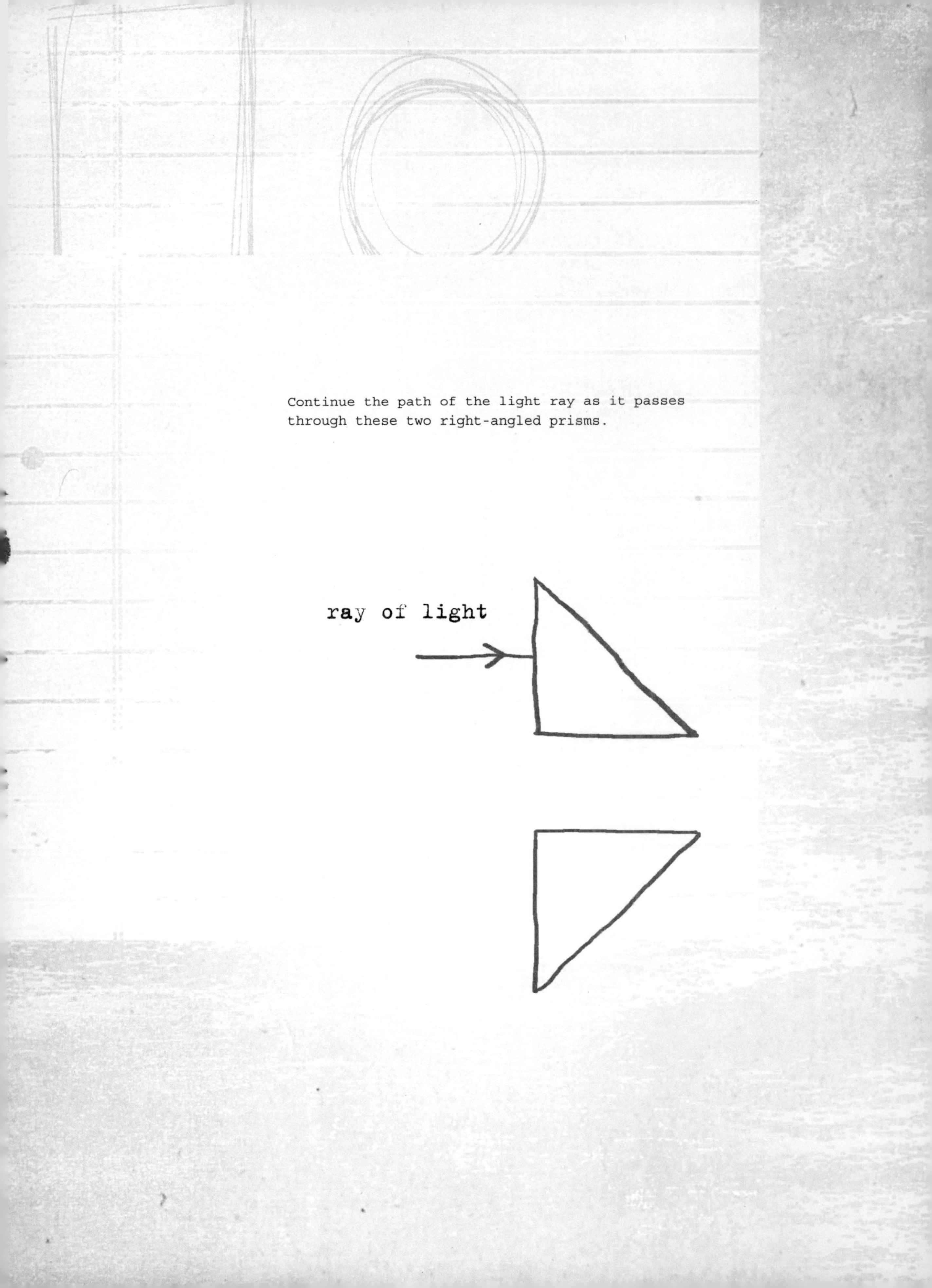

A railway wagon of mass 5,000 kilograms is moving at 1.5 m/s. It couples with a stationary second wagon of 8,000 kilograms. What is the velocity of the two wagons after they have collided?

answers p126/7

p6

Red, orange, yellow, green, blue, indigo, violet

p7

The force of gravity always acts toward the center of the Earth, wherever you are on it, so all four arrows should point toward the center of the Earth

p8

1) Protons have a positive charge
2) Electrons have a negative charge
3) Neutrons have no charge

p9

Opposite poles attract and like poles repel, so magnet A should be labeled N S and magnet C should be labeled S N

p10

It moves lower (the heat causes the metal wire to expand)

p11

The shadow cast would be a circle on the screen.

p12

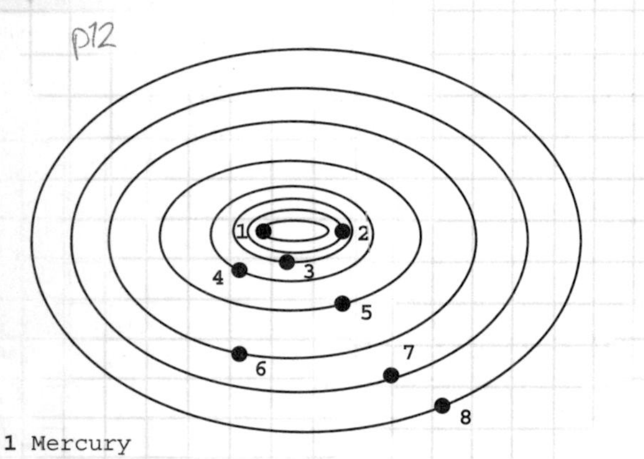

1 Mercury
2 Venus
3 Earth
4 Mars
5 Jupiter
6 Saturn
7 Uranus
8 Neptune

p13

It would attract more paperclips (the strength of an electromagnet increases as the number of coils in the wire increases)

p14

1) All three bells will ring (the circuit is completely closed so the current flows through all components)
2) Only bells Y and Z will ring (the current is broken at switch A so will not reach bell X)

p15

1) Mercury
2) Neptune (the further away the planet the more slowly it moves and the longer it takes to orbit the Sun)

p16

Resistance always acts in the opposite direction to the movement of an object, so in this case the arrow should point upwards.

p17

The clearer the liquid, the more light units recorded, so:
Water = C
Cooking oil = B
Blackcurrant squash = A
Thick black coffee = D

p18

Opposite poles attract and like poles repel, so magnet Z will move magnet X along the table in the direction of the arrow; magnet Y will be repelled by magnet X and will fall off the table, because the south poles are facing each other.

p19

The ray should be a straight line from the cyclist to the glass of the shop window and then from the glass to the motorcyclist. The angles should be equal.

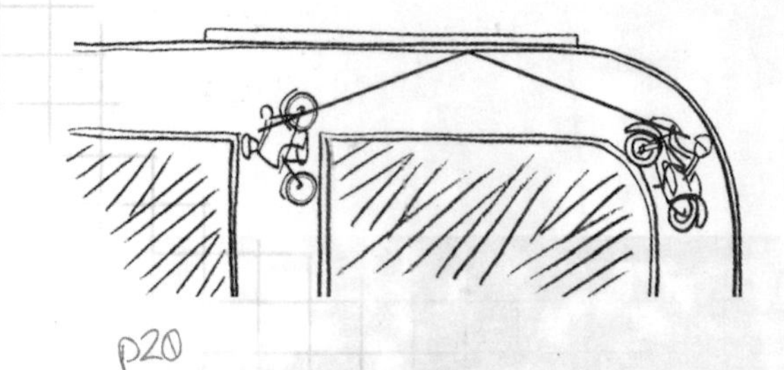

p20

The light source should be placed at A.

p21

C (the effect of the Sun's gravity is greatest when the asteroid is closest to it)

P22

2) Magnet D is stronger than magnet B.
(It attracted more paperclips from the same distance away.)

p23

The wrench and the feather hit the surface at the same time (there is no air resistance on the Moon to oppose the force of gravity)

p24

The circle in P and the square in R would both be blocked by the cardboard in Q, so neither shape would appear on the screen. You would only see the triangle and the shadow of the cardboard:

p25

A is a series circuit (connected so the current passes through all components)
B is a parallel circuit (connected so the current divides before recombining)

p26

North pole
Opposite poles are touching, so if A = N, then B = S, C = N, D = S, G = N, H = S, so J = N

p27

C) Your weight and the reaction force from the ground (as you are not moving up and down, these forces must be balanced)

p28

1.8 feet per second
Average speed = distance traveled (feet) ÷ time taken (seconds) so 18 ÷ 10 = 1.8

p29

Rollercoaster = gravitational potential energy
Guitar string = sound energy
Coffee = thermal energy
Battery = chemical energy
Parachutist = kinetic energy

p30

Spring balance C (the fulcrum is farthest from the load, so a greater force will be needed to balance it)

p31

A is off (the circuit is broken) and B is on (the circuit is complete)

p32

Kinetic energy (energy of movement, which makes the blades turn)

p33

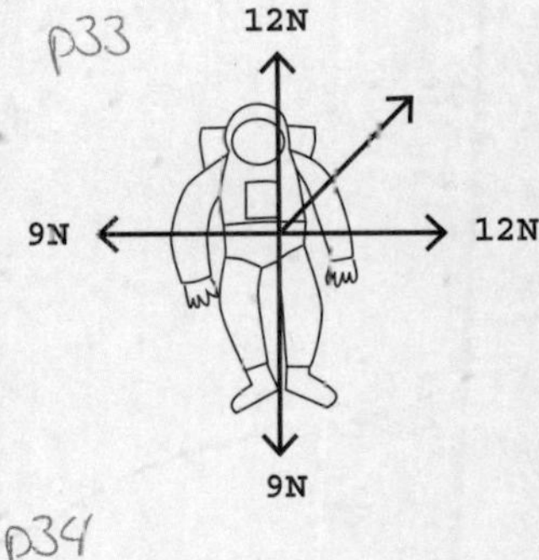

The astronaut will move in the direction of the midpoint of the two equal strongest forces:

p34

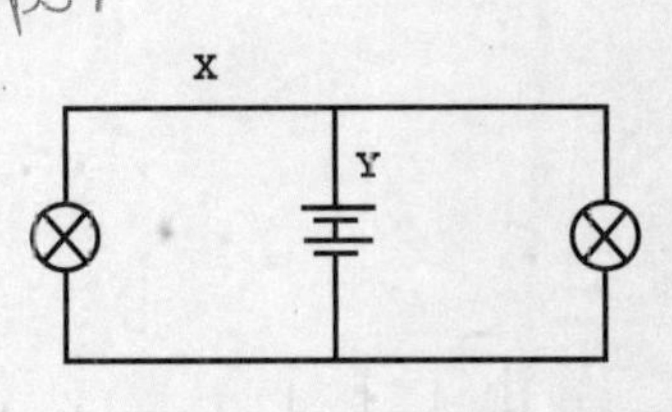

To light the front lamp only, the switch must be placed after the current divides. To light both lamps it must be before the current divides.

p35

4 inches
Work out how much the spring extends for every lb added: 6 lb − 2 lb = 4 lb; 13 inches − 7 inches = 6 inches, so for 4 lb extra weight, the spring extends 6 inches; 6 ÷ 4 = 1.5 (verify this by doing the same calculation with springs B and C).
If spring A measures 7 inches with 2 lb weight, then with 0 lb weight it will measure 4 inches (1.5 x 2 = 3; 7 − 3 = 4)

p36

1) Clockwise (if gears are connected by a chain or belt, they move in the same direction)
2) 4 turns (if gear A has 6 teeth and makes 8 turns, gear B—with twice as many teeth—will make half the number of turns)

p37

The circuit diagram should show a single battery and three lamps—two in series and one in parallel with the other two.

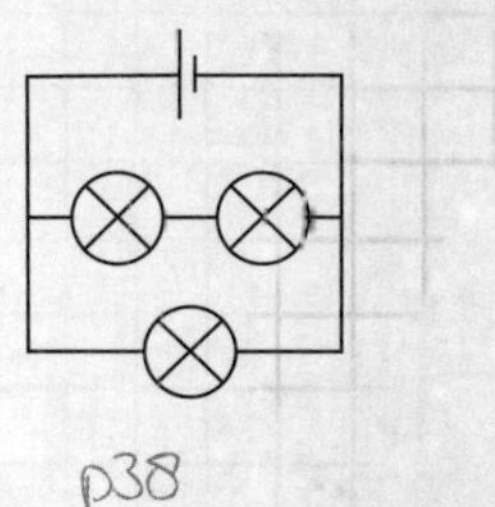

p38

Jupiter (the greater the force of gravity on the planet, the greater the counter force will need to be for the spaceship to lift off)

p39

240 Nm
To find the turning moment (measured in newton-meters), multiply the height by the force:
0.8 x 300 = 240

p40

The comet travels fastest when it is closest to the Sun. A body's angular momentum (the mass of a body multiplied by its velocity and the radius of its orbit) must remain the same at all points. As its mass remains constant, a body moving in an elliptical orbit must experience increased velocity closest to the Sun because the radius is smaller.

p41

The ray will reflect off the mirror at an equal angle to the incident ray

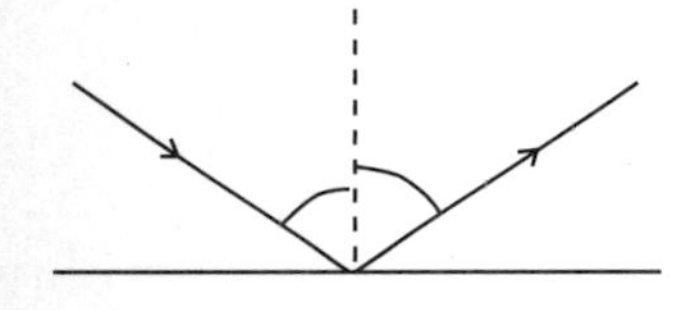

p42

The lamps would be dimmer if the batteries were arranged in parallel:

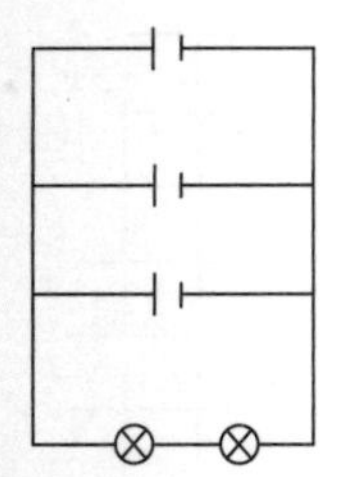

p43

A = Lift
B = Air resistance (drag)
C = Gravity
D = Thrust

p44

C) West to east

p45

Rods X and Y conduct electricity
Lamp 1 will only be lit if there is a conductor at points A and C
Lamp 2 will only be lit if there is a conductor at points B and C
Lamp 3 will only be lit if there is a conductor at either A or B, and C

p46

A = wavelength
B = amplitude
C = crest
D = trough
E = displacement

p47

gravitational potential energy → kinetic energy → kinetic energy → electrical energy

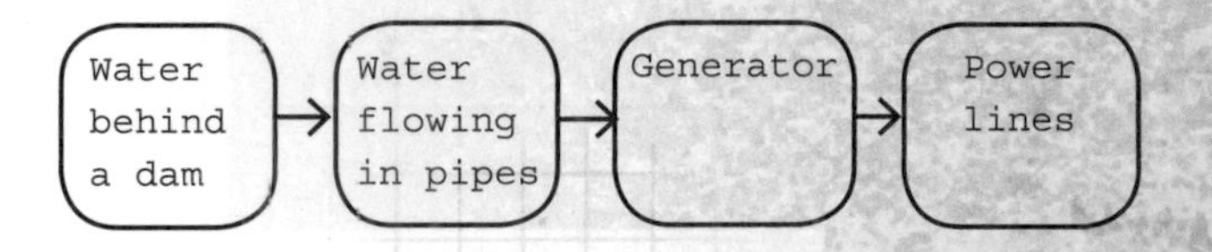

p48

11 million newtons (resultant force = upward force - weight; 27 - 16 = 11)

p49

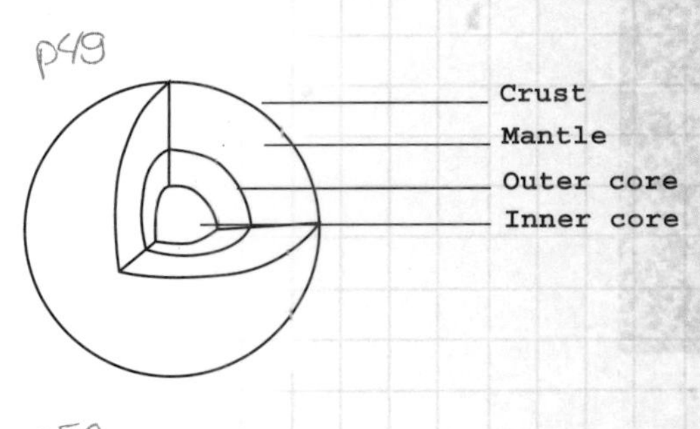

p50

Counterclockwise at 5 rpm
If gears are touching, then adjacent gears move in opposite directions. Gears with the same number of teeth move at the same speed, but speed is inversely proportional to the number of teeth. If gear X has 10 teeth and moves at 10 rpm, gear Y, with 20 teeth, will move at half that speed.

p51

The louder the sound, the greater the amplitude of the wave; the higher the pitch, the greater the frequency of the wave, so:
C is a quiet, high-pitched sound
B is a loud, low-pitched sound
A is a loud, high-pitched sound

p52

None (force = mass x acceleration; constant velocity means there is no acceleration so there is no net force—this is Newton's first law of motion, or the law of inertia)

p53

10 newtons
The plank is balanced, so the clockwise moment must equal the counterclockwise moment.
Find the clockwise moment: 0.5 x 5 = 2.5 Nm
The counterclockwise moment must also be 2.5 Nm;
F x 0.25 = 2.5; F = 2.5 ÷ 0.25 = 10

p54

The ray bends to the right as it enters the prism, and then will bend downwards as it leaves the prism:

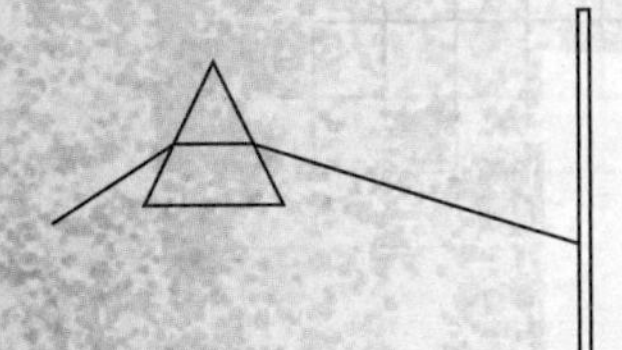

p55

Direct current always flows in the same direction so is marked by a straight line as shown. Alternating current changes direction at regular intervals, so would be represented on a graph as:

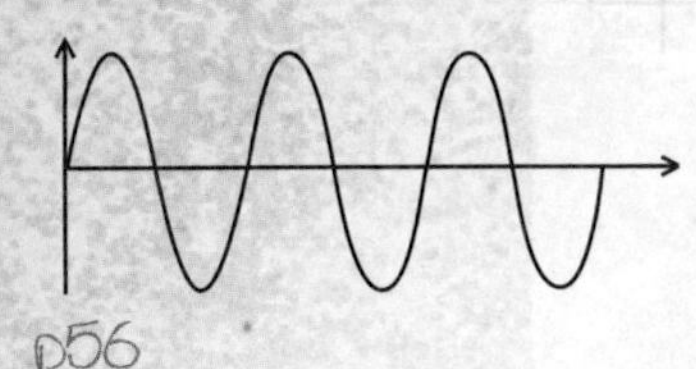

p56

Pressure is calculated as force per unit area: force ÷ area, so:
1) 100 newtons per square inch (200 ÷ 2 = 100)
2) 800 newtons (100 x 8 = 800)

p57

10,200 meters (distance = speed x time;
340 x 30 = 10,200)

p58

Moment = force x distance, so:
1) 45 Nm (15 x 3)
2) 45 Nm (30 x 1.5)

p59

1.5 amps (current in a circuit depends on the number of cells; adding more cells increases the current, so if one cells provides 0.5 amps, then three cells will provide 1.5 amps)

p60

Machine X requires the least effort (the effort is applied at A, while the load is applied at B; A has a proportionally larger radius than B, so less effort is required)

p61

A total eclipse takes place in the area where the Moon passes directly in front of the Sun and appears to cover it completely, as viewed from Earth; a partial eclipse is viewed elsewhere:

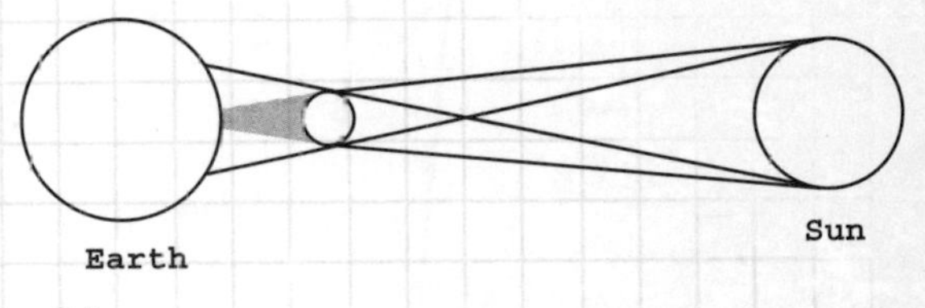

p62

Frequency = 10 Hz (there are 10 waves each second)
Speed = 7.5 feet per second (distance ÷ time; 30 ÷ 4 = 7.5)
Wavelength = 0.75 feet (speed ÷ frequency; 7.5 ÷ 10 = 0.75)

p63

80 kilograms (mass doesn't vary with location, so the astronaut's mass will be the same on the Moon as it is on Earth)

p64

They should all be labeled 3 amps (the current is the same everywhere in a series circuit)

p65

The magnetic field around a straight wire is circular and at right angles to the wire.

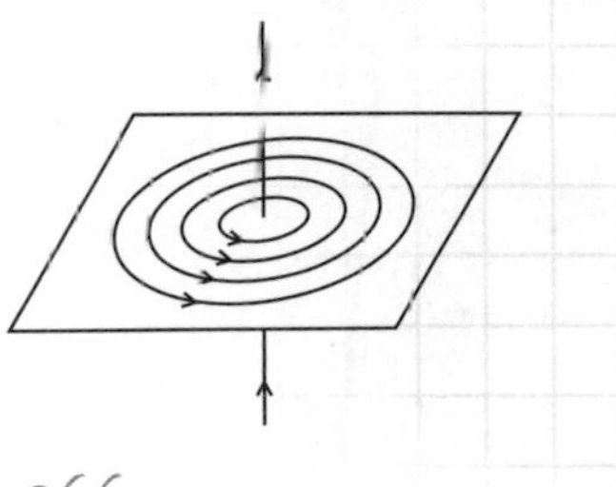

p66

From left to right (longest to shortest wavelength): radio waves, infrared, ultraviolet, gamma rays

p67

1) A and F (kinetic energy is energy of movement; at these two points the car is not moving so it has no kinetic energy)
2) A (gravitational potential energy is stored energy, so the car has most at the top of the ride)
3) Gravity (the car moves down the slope by its weight or gravitational force)

p68

1.3 miles
distance = speed x time; 0.2 x 6.5 = 1.3

p69

1) The angle of incidence is always greater than the angle of refraction
2) 30°

p70

B (a pendulum has most kinetic energy when all its gravitational potential energy has been converted to kinetic energy, i.e. at the straight vertical position)

p71

8 minutes and 19 seconds
Find the average distance of the Sun from Earth: 91,402,000 + 94,512,000 = 185,914,000; 185,914,000 ÷ 2 = 92,957,000
Divide the average distance by the time: 92,957,000 ÷ 186,282 = 499 seconds
499 = 8 minutes and 19 seconds

p72

Sound waves are compressed as a source approaches an observer, so the wavelength **decreases** and frequency **increases**.
Sound waves are stretched as a source moves away from an observer, so the wavelength **increases** and frequency **decreases**.

p73

1,300 newtons (weight = mass x gravity; 50 x 26 = 1,300]

p74

All of the lamps are lit (the copper wire keeps all the lamps connected in the circuit)

p75

1) Unbalanced (the speed of the car is decreasing as it brakes, so the forces acting on it are not balanced)
2) Unbalanced (although speed is constant, the satellite's direction is changing all the time, so the forces are not balanced)
3) Balanced (the speed and the direction of the cyclist are both constant, so the forces are balanced)

p76

A) The temperature stays the same (temperature of a substance remains the same during melting, boiling, or freezing)

p77

1) 3.7 times larger (1 ÷ 0.27 = 3.7)
2) 13.3 times farther away (5.2 ÷ 0.39 = 13.3)

p78

1) If the force is doubled, the acceleration would double (force = mass x acceleration, so if the force is doubled and the mass stays the same, acceleration must increase proportionally)
2) If the mass of the trolley is doubled, the acceleration would halve (using the same logic as above, if the force remains the same but the mass doubles, then the acceleration must halve)

p79

1)

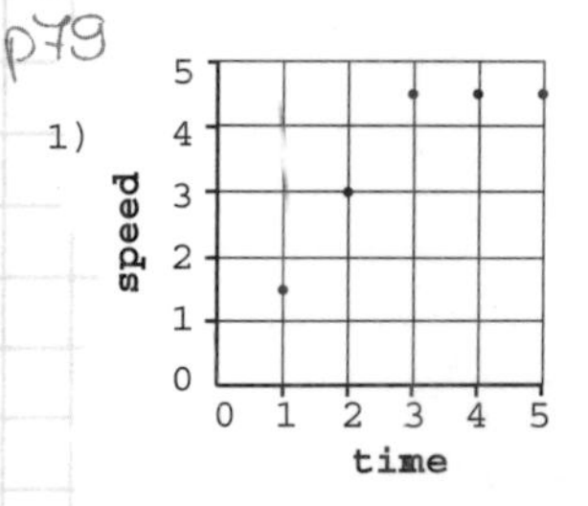

2) 1.5 m/s² (acceleration = change in speed ÷ time; 4.5 ÷ 3 = 1.5)

p80

1 joule is 1 watt x no. of seconds, so multiply the power rating in watts by the time in seconds.
1 kWh is 3,600,000 joules (a kilowatt is 1,000 watts and an hour is 3,600 seconds), so divide the energy in joules by 3,600,000.

Appliance	Energy in joules	Energy (kWh)
Kettle	1,200,000 (2,000 x 600)	0.33
Iron	2,160,000 (1,200 x 1,800)	0.6
Vacuum cleaner	1,200,000 (1,000 x 1,200)	0.33

p81

75 Nm (turning moment = force x distance; 50 x 1.5 = 75)

p82

20 years
Payback time = initial cost ÷ annual saving; 7,000 ÷ 350 = 20

p83

50 feet
Distance traveled = speed of ultrasound x time delay
A millisecond is 1,000th of a second
Depth of the water is half the distance traveled by the pulse (it has to go there and back)
5,000 x (20 ÷ 1,000) ÷ 2 = 50

p84

300 m/s^2 (acceleration = change in speed ÷ time taken; 150 ÷ 0.5 = 300)

p85

25 joules of heat energy

All light bulbs transfer some energy as wasted heat.

Sankey diagrams show the types and proportions of energy in a transfer, so if light energy is 75 joules, then the rest must be 25 joules (no energy is lost, it is just transferred from one form to another)

p86

7 volts

Potential difference (measured in volts) = current x resistance; 2.0 x 3.5 = 7.0

p87

Device	Energy produced	Digital/analog
Loudspeaker	Sound	Analog
Electric motor	Kinetic	Analog
Light-emitting diode (LED)	Light	Digital

p88

33 cents

Convert watts to kilowatts: 2,000 ÷ 1,000 = 2

Work out the kilowatt-hours: 2 x 0.75 = 1.5 (don't multiply by 45, convert to hours—45 minutes is 0.75 hours)

Multiply kilowatt-hours by cost: 1.5 x 22 = 33

p89

Amplitude would decrease and frequency would increase:

p90

1.45 m/s^2 (acceleration = net force ÷ mass; 80 ÷ 55 = 1.45)

p91

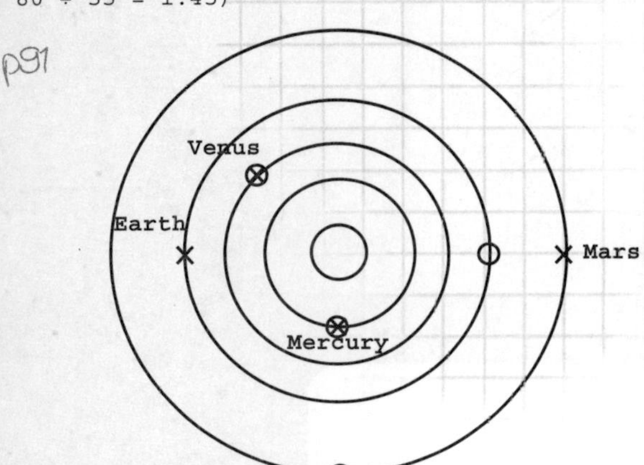

p92

270 coulombs

Charge = current x time; time must be converted to seconds; 1.5 x (3 x 60) = 270

p93

1) Car A has the greatest acceleration because the gradient of the line is greater (gradient is equal to acceleration on a velocity-time graph)

2) Car B has traveled farthest because the area under the line for car B is greater than that of car A (it's not just that the line for car B is longer)

p94

0.025 meters

Wavelength = speed ÷ frequency

Convert the speed from km/s to m/s: 2 x 1,000 = 2,000

2,000 ÷ 80,000 = 0.025

p95

1) Gamma (gamma rays are waves not particles, at the short-wavelength end of the EM spectrum)

2) Alpha (alpha particles are made up of 2 protons and 2 neutrons, which is a helium nucleus)

3) Beta (beta particles have a single negative charge—they are electrons)

p96

15 meters

Distance traveled = velocity x time, so at 20 m/s the distance traveled is 10 meters (20 x 0.5 seconds).

To find the thinking distance at 30 m/s, find the difference in speed (30 ÷ 20 = 1.5) and multiply by the distance at 20 m/s (10 x 1.5 = 15)

p97

The waves are diffracted (spread out) so they curve; however wavelength (distance between each wave front) remains the same.

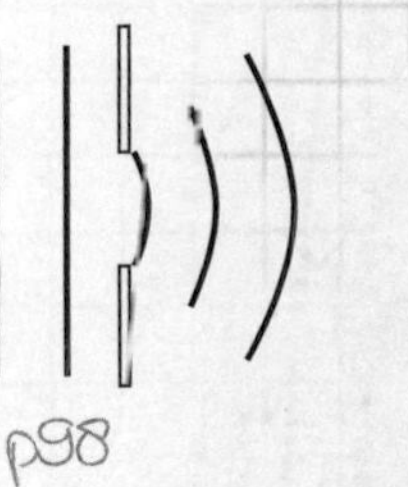

p98

1) Electrons

2) A negative charge (electrons carry a negative charge, so in gaining electrons, the balloon has gained a negative charge; it can therefore stick to the positively charged surface of the wall)

p99

1) 150 meters (distance = speed x time; 30 x 5 = 150) 2) 10 m/s^2 (deceleration = change in velocity ÷ time; 30 ÷ 3 = 10)

p100

A and C (isotopes of an element have the same number of protons and electrons, but a different number of neutrons)

p101

7.5 ohms (resistance = voltage ÷ current; 3 ÷ 0.4 = 7.5)

p102

B is the digital wave (because digital signals consist of only two states, on or off, there are no values in between)

p103

37.5 joules
Kinetic energy = ½ mass x velocity^2;
(12 ÷ 2) x 2.5^2; 6 x 6.25 = 37.5 joules

p104

5 meters
Wavelength = speed ÷ frequency; 3 x 10^8 ÷ 60 x 10^6 = 5
For the bonus point: its long wavelength tells you it's a radio wave

p105

Average time is 0.578 seconds (0.55 + 0. 62 + 0.58 + 0.56 + 0.58 ÷ 5)
Speed of sound is 346 m/s (distance ÷ time; 200 ÷ 0.578 = 346)

p106

2.3 kilowatts (power = current x voltage; 10 x 230 = 2,300 watts; to convert from watts to kilowatts, divide by 1,000)

p107

500 newtons forwards Horizontal resultant force = thrust - drag; 2,000 - 1,500 = 500

p108

1) 30 meters west (displacement is the distance and direction you are from your starting point)
2) 2.6 m/s (speed = distance ÷ time; 130 ÷ 50 = 2.6)

p109

20 Hz (the normal range of human hearing is 20 Hz to 20,000 Hz; sounds with frequencies lower than 20 Hz are infrasound; sounds with frequencies higher than 20,000 are ultrasound)

p110

55.6 joules (10 ÷ 0.18 = 5.6)

p111

1) 2 seconds (after that the graph is negative because the ball is falling back to Earth)
2) 20 meters (height = area under the graph; ½ x 20 x 2 = 20)

p112

1.8 volts (voltage = current x resistance; 3 x 0.6 = 1.8)

p113

1) 680 m/s (speed = distance ÷ time; convert kilometers to meters; 1,020 ÷ 1.5 = 680)
2) 340 m/s (the airplane is traveling at twice the speed of sound; 680 ÷ 2 = 340)

p114

6,000 newtons
force x distance = ½ mass x velocity^2
F x 75 = ½ x 1,000 x 30^2
F x 75 = 1,000 x 900 ÷ 2
F x 75 = 450,000; F = 450,000 ÷ 75 = 6,000

p115

2.5 amps (current = voltage ÷ resistance; 10 ÷ 4 = 2.5)

p116

156 seconds
Half-life is the time it takes for the activity to halve: 640 - 320 - 160 - 80 = 80, so it takes three half-lives to reach 80 counts per second. To find the number of seconds, multiply the number of seconds per half-life (52) with the number of half-lives (3); 52 x 3 = 156

p117

5,000 newtons
Calculate kinetic energy = 0.5 x 800 x 25^2 = 250,000 joules
Force = energy ÷ distance = 250,000 ÷ 50 = 5,000

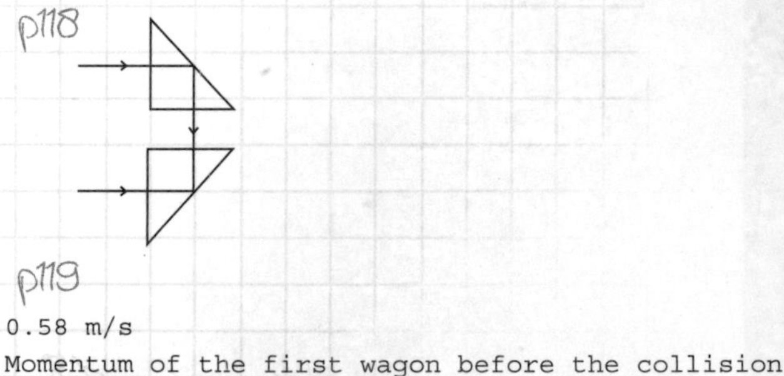

p119

0.58 m/s
Momentum of the first wagon before the collision = 5,000 x 1.5 = 7,500 kg m/s
Momentum of the second wagon before the collision = 0
Momentum remains the same before and after the collision, so momentum after = 7,500 kg m/s
Mass after collision = 5,000 + 8,000 = 13,000 kg
7,500 ÷ 13,000 = 0.58

USEFUL EQUATIONS

acceleration = change in velocity ÷ time

charge = current x time in seconds

efficiency = (useful energy transferred ÷ energy supplied) x 100%

electrical power = current x voltage

force = mass x acceleration

kinetic energy = ½ mass x velocity^2

moment = force x distance

pressure = force ÷ area

resultant force = the difference between two forces acting on an object

speed = distance ÷ time

voltage = current x resistance

wavelength = speed ÷ frequency

weight = mass x gravity

work done = force x distance moved

Thunder Bay Press
An imprint of the Baker & Taylor Publishing Group
10350 Barnes Canyon Road, San Diego, CA 92121
www.thunderbaybooks.com

All notations of errors or omissions should be addressed to Thunder Bay Press, Editorial Department, at the above address. All other correspondence (author inquiries, permissions) concerning the content of this book should be addressed to Paperwasp at the address below.

This book was conceived, designed, and produced by Paperwasp, an imprint of Balley Design Limited, The Mews, 16 Wilbury Grove, Hove, East Sussex, BN3 3JQ, UK
www.paperwaspbooks.com.

Creative director: Simon Balley
Designer: Kevin Knight
Project editor: Sonya Newland
Illustrations: Kevin Knight

ISBN-13: 978-1-60710-439-1
ISBN-10: 1-60710-439-3

Printed in China.

1 2 3 4 5 16 15 14 13 12